Leadership Approaches to the Science of Water and Sustainability

Kristin Joyce Tardif
University of Arkansas, Fort Smith, USA

A volume in the Practice, Progress, and Proficiency in Sustainability (PPPS) Book Series

IGI Global
PUBLISHER of TIMELY KNOWLEDGE

Published in the United States of America by
IGI Global
Engineering Science Reference (an imprint of IGI Global)
701 E. Chocolate Avenue
Hershey PA, USA 17033
Tel: 717-533-8845
Fax: 717-533-8661
E-mail: cust@igi-global.com
Web site: http://www.igi-global.com

Library of Congress Cataloging-in-Publication Data

Names: Tardif, Kristin Joyce, 1955- author.
Title: Leadership approaches to the science of water and sustainability / by Kristin Joyce Tardif.
Description: Hershey PA : Engineering Science Reference, [2022] | Includes bibliographical references and index. | Summary: "This book covers a brief geologic history of water, a collection of memoirs of exploring and looking for pristine spring water sources, about the complications of quality control, water law and the science of water and sustainability"-- Provided by publisher.
Identifiers: LCCN 2021051165 (print) | LCCN 2021051166 (ebook) | ISBN 9781799896913 (hardcover) | ISBN 9781799896920 (paperback) | ISBN 9781799896937 (ebook)
Subjects: LCSH: Tardif, Kristin Joyce, 1955- | Hydrology--United States. | Water resources development--United States. | Water--Law and legislation--United States. | Springs--United States. | Leadership. | Women earth scientists--United States--Biography.
Classification: LCC GB661.2 .T37 2022 (print) | LCC GB661.2 (ebook) | DDC 333.91/17--dc23/eng/20211130
LC record available at https://lccn.loc.gov/2021051165
LC ebook record available at https://lccn.loc.gov/2021051166

This book is published in the IGI Global book series Practice, Progress, and Proficiency in Sustainability (PPPS) (ISSN: 2330-3271; eISSN: 2330-328X)

British Cataloguing in Publication Data
A Cataloguing in Publication record for this book is available from the British Library.

Practice, Progress, and Proficiency in Sustainability (PPPS) Book Series

ISSN:2330-3271
EISSN:2330-328X

MISSION

In a world where traditional business practices are reconsidered and economic activity is performed in a global context, new areas of economic developments are recognized as the key enablers of wealth and income production. This knowledge of information technologies provides infrastructures, systems, and services towards sustainable development.

The **Practices, Progress, and Proficiency in Sustainability (PPPS) Book Series** focuses on the local and global challenges, business opportunities, and societal needs surrounding international collaboration and sustainable development of technology. This series brings together academics, researchers, entrepreneurs, policy makers and government officers aiming to contribute to the progress and proficiency in sustainability.

COVERAGE

- Knowledge clusters
- Innovation Networks
- Intellectual Capital
- Socio-Economic
- Global Business
- E-Development
- Environmental informatics
- Outsourcing
- Green Technology
- Eco-Innovation

The Practice, Progress, and Proficiency in Sustainability (PPPS) Book Series (ISSN 2330-3271) is published by IGI Global, 701 E. Chocolate Avenue, Hershey, PA 17033-1240, USA, www.igi-global.com. This series is composed of titles available for purchase individually; each title is edited to be contextually exclusive from any other title within the series. For pricing and ordering information please visit http://www.igi-global.com/book-series/practice-progress-proficiency-sustainability/73810. Postmaster: Send all address changes to above address.

Titles in this Series

For a list of additional titles in this series, please visit:
http://www.igi-global.com/book-series/practice-progress-proficiency-sustainability/73810

Climatic and Environmental Significance of Wetlands Case Studies from Eurasia and North Africa
Abdelkrim Ben Salem (Mohammed V University, Morocco) Laila Rhazi (Mohammed V University, Morocco) and Ahmed Karmaoui (University of Moulay Ismail, Morocco & Southern Center of Culture and Science, Morocco)
Engineering Science Reference • © 2022 • 310pp • H/C (ISBN: 9781799892892) • US $250.00

Eco-Friendly and Agile Energy Strategies and Policy Development
Mir Sayed Shah Danish (University of the Ryukyus, Japan) and Tomonobu Senjyu (University of the Ryukyus, Japan)
Engineering Science Reference • © 2022 • 335pp • H/C (ISBN: 9781799895022) • US $240.00

Handbook of Research on Global Institutional Roles for Inclusive Development
Neeta Baporikar (Namibia University of Science and Technology, Namibia & University of Pune, India)
Information Science Reference • © 2022 • 398pp • H/C (ISBN: 9781668424483) • US $270.00

Analyzing Sustainability in Peripheral, Ultra-Peripheral, and Low-Density Regions
Rui Alexandre Castanho (WSB University, Poland)
Engineering Science Reference • © 2022 • 334pp • H/C (ISBN: 9781668445488) • US $240.00

Frameworks for Sustainable Development Goals to Manage Economic, Social, and Environmental Shocks and Disasters
Cristina Raluca Gh. Popescu (University of Bucharest, Romania & The Bucharest University of Economic Studies, Romania)
Engineering Science Reference • © 2022 • 296pp • H/C (ISBN: 9781668467503) • US $240.00

IGI Global
PUBLISHER of TIMELY KNOWLEDGE

701 East Chocolate Avenue, Hershey, PA 17033, USA
Tel: 717-533-8845 x100 • Fax: 717-533-8661
E-Mail: cust@igi-global.com • www.igi-global.com

Table of Contents

Foreword

My name is Dr. Ken Warden. I am the previous Dean of the once College of Applied Science and Technology at the University of Arkansas-Fort Smith. I have spent the past 24 years in higher education as a faculty member and administrator in both two-year and four institutions in Arkansas.

I first met Dr. Kristin Tardif in the same timeframe I was interviewing for a new position at the University of Arkansas Fort Smith in 2014. She had a bright personality and an inviting smile. I didn't realize at the time she was a faculty member in the college to which I was applying to be dean, and she didn't know I was applying. We immediately made a connection that would evolve into a supportive professional network and great friendship.

As I learned more about her, I learned of her extensive prior work in the corporate world working for a multi-national company in the area of water and sustainable resources. I can only imagine how difficult it must have been raising four children while climbing a corporate structure to become one of the leaders in her organization. It was a different time in the '80s. A woman navigating corporate America at that time was much different and much more difficult than I think it is today. It's women like Dr. Kristin Tardif that broke the glass ceilings and helped pave the road for those behind her. Blazing trails and lifting others as she goes is an identifying characteristic of Kristin Tardif.

In her second career, she came to education. Paying out-of-pocket to study her passion and investing monetarily in something she knew would arguably take too many years to realize any "R.O.I". But to people like Dr. Tardif, the greatest R.O.I. is not measured monetarily. She earned her doctorate in … Organizational Leadership from The Chicago School of Professional Psychology in 2015. She was the first person I knew that truly understood "Leadership" as its own unique discipline. It's not management and it's not business administration, these are technical skills others profess. Dr. Tardif professes leadership, and she does a damn fine job of it.

As a dean responsible for programs outside of my past experiences, I initially came to the knowledge of leadership, as a field of study, through conversations with Dr. Tardif. It made a lot of sense to me. As she led the program at UA Fort Smith, I watched her put her academic knowledge into skill practice. She mentored students and faculty alike. She said "we" when things went well and "I" when things went wrong. She used soft hands when needed and stern hands when required. She always knows which hands to use.

I have watched her invest in students over and over and over again, knowing they didn't necessarily deserve it, but that they needed it to thrive and be their best selves. She has pulled more students across the finish line, some kicking and screaming than any other faculty member I have ever known. She is a teacher and a scholar. Further, she has a heart for service that is second to none. She is one of the hardest-working people I have ever known.

In this book, Leadership Approaches to the Science of Water and Sustainability, Dr. Tardif chronicles some precarious activities and some hilarious circumstances, all the while pulling the reader into the science of water and sustainability. Storytelling is an art, an art that has existed for thousands of years. Stories bring merit to concepts beyond what facts and figures only can often bring. When told correctly, stories add tremendous color and relevance to otherwise dull topics. The ability to tell stories and shine the light on relevance is one of the key components in the make-up of a great teacher. When topics are mundane, the best faculty members use past experiences and stories to articulate the relevance to students; make the learning fun while ultimately transferring knowledge in a way that ensures the learner understands, and more importantly, retains the content.

This book is a prime example of how stories can make the science fun and relevant while helping the learner retain and grow. I hope you enjoy this book as much as I. Laugh with Dr. Tardif and enjoy learning the science.

Kenneth R. Warden
Independent Researcher, USA

Preface

Autoethnography is used in the practice of adult education. Often called *Voices from the Field.* Autoethnography narratives reveal experiences in the workplace, research settings, community organizations, and other contexts. This is the systematic analysis of (graphy) personal experience (auto) in order to understand cultural experience (ethno) (Ellis et al, 2011). This took connects theory and practice. Autoethnography accounts provide readers with narratives that add to the knowledge of the experiences of the learner, instructors, and researchers in adult education.

Personal Reflections are thoughtful reflections on and discussion of important learning and teaching issues in adult education. First-person narratives are thought-provoking experiences in teaching adults or examinations of taken-for-granted practices about adult learning, teaching, program design, and management in adult education. Personal Reflection examines the usefulness of autoethnography in the field of adult education.

This book is the personal reflections of a scientist in the field, science of water history and science of water. These personal experiences reflect the need for water education and knowledge of environmental laws. Each reflection details how little knowledge is known of basic water science and water law. In these experiences the scientist has met the best of life, some she questioned, and some that were just naïve but trying to make a living and better themselves. Some were just not right! These stories are true, some endearing and some funny. Moreover, they are science lessons on saving the greatest resource for mankind.

The first part of the book is about the science of water history in the U.S., and the search for pristine sources. After an exhaustive search, finding a source, hours were spent a state and local court houses and combing local archives to find the land uses going back in some cases to the 1600s. I was able to see some of the most beautiful places on the planet, ride in helicopters over the White Mountains, kayak down pristine rivers, and hike in the forest primeval

By beginning the book with the history of water and sustainability in the U.S. we Mankind did r understand the result of their actions, many were just trying to survive the harsh environment in a new world. Many mistakes were made, and habitat and land management were not even in someone's purview. By going through history, we understand how water rights and water rights laws were established. The second chapter is a detailed description of the water rights law, which is vastly different in areas of the U.S. and the world.

The next four chapters focus on memoirs and the science of land sustainability and resource management. In the memoirs, the spring owners did not fully understand what can happen to a water resource when strip mining is nearby and overgrazing can alter and contaminate beautiful water resources. Today, we have to look at some of the green energy and study if they are really green. Lithium is mined by strip mining and it destroys the environment and groundwater near the mine. In Chapter 7, we learn how nature is fragile and mighty. A lawyer or company cannot alter a bog or wetland because it is inconvenient.

Chapter 9 is near and dear to my heart because I worked in Canada for many years and realized how different the U.S. Canada, and the world view water and varied the desired taste for water can be. Often, we only look at our own view which is very limited. Chapter 10 is a salute to those scientists who brave isolated, cold, hot conditions to devote life-long research to understand and protect our planet. I followed this chapter with Chapter 11, "The Squatters". Squatting is becoming a global phenomenon. This puts our forest and natural resources at risk due to how squatters live off the land, but we have to really address why this started and as a society what we should be doing to take care of each other better.

The next two chapters go over a class action lawsuit, and quality control. Most companies are heavily regulated, especially if they are large companies. My hope for this chapter is to give the reader and researcher a different point of view. Lawyers fly in private jets and make millions drumming up an idea to file lawsuits. In addition, we have somehow, trying to protect customers developed a path of ethical behavior.

Chapter 14 is on solutions. How can the creative economy help us create a sustainable world? Every great new idea, service, or product came from creativity and innovation. It did not come from politics. Politics has had the tendency to destroy creativity. We have to lead by science, creativity, and innovation. The final chapter is a glossary, this helps the reader understand the terminology used in the book. This book uses autoethnography, with the stories demonstrating how the application or not, of the theories of science, law, and policy shape our world. Most importantly, how creativity and innovation are our future will bring about a better world to live in.

Kristin Joyce Tardif
University of Arkansas, Fort Smith, USA

Introduction

This book is about a part of my life that few people experience. I have met the best of life, some I questioned, and some that were just naïve but trying to make a living and better themselves. Some were just not right! These memoirs are true, I hope you find some endearing, some funny and enjoy the madness found in strange places. Names and places have not been named or changed to protect people's privacy.

In addition, this a geologic history of water, a collection of memoirs of exploring and looking for pristine spring water source, about the complications of quality control, water law and the science of water. After an exhaustive search, finding a source, hours were spent a state and local court houses and combing local archives to find the land uses going back in some cases to the 1600's. I met good people with a love of nature, good people that ruined the land without knowing it, some characters, and some I would like to forget. I kept those out of the book. I was able to see some of the most beautiful placers on the planet, ride in helicopters over mountains, and kayak down pristine rivers, and hike in the forest primeval.

I wish I would have logged the hours I walked; it was definitely easier to keep weight off in those days. I do not miss coming home tired, dirty and trying to get ticks off of me. Sometimes over sixty. Many times, I work in extreme weather conditions, one time being -38 degrees, this was not the windchill. I thank the volunteer, a contractor, that out of the goodness in his heart, came with me, with a snow mobile. If he had not been so kind, I would have been on snow shoes, and probably would not have finished in a day. A big thank you to the kind Fed Ex person, who gave me hot coffee while she labeled, packed my water samples and mailed them to laboratories all over the U.S. The kindness of people always surprises me but blesses me. I actually made in home to cook supper for my children by 7:30pm. For the warriors looking for good resources and those trying to save, them the path is not always easy but arduous.

AUTOETHNOGRAPHY

The methodology for this book is based on autoethnography. Autoethnography is an approach to researching and writing with description and systematic analysis of personal experience to understand science and cultural experience. The challenges of this approach are not to be autobiographical. This method is both a process and a product. It challenges the canonical ways of doing research and representing others (Ellis, Adams & Boucher, 2011).

Autoethnography is an emerging qualitative research method that allows the author to write in a highly personalized style, drawing on their experience to extend understanding about a phenomenon or science. Autoethnography is grounded in postmodern philosophy and is linked to growing debate about voice in research. The intent of autoethnography is to acknowledge the link between the personal and the cultural, to have a better understanding of inquiry and expression (Wall, 2006).

Scholars working across disciplines have now considered social science akin to physics, by telling stories with theories, they centered rather than pretending to be of value. In addition, scholars have turned to autoethnography seeking a positive response to critiques of canonical ideas about what research is and how it should be done. By focusing on producing meaningful, accessible and evocative research grounded in personal experience, research that would sensitize the reader to many complex issues. Autoethnographers understand the ways personal experience influences the process (Ellis, Adams & Boucher, 2011).

Normally, a researcher decides who, what, when where and how to research decisions that are tied to institutional requirements. Moreover, researchers will still often will change names and places for protection. Many researchers now recognize that it is difficult to be objective. Autoethnography is one of the approaches that acknowledges and accommodates subjectivity, emotionality, and the researcher's influence on research, rather than hiding from these matters or assuming they don't exist. In addition, many scholars have been recognizing that there are multi- cultures, and many ways of expressing ourselves (Ellis, Adams, & Boucher, 2011).

For myself, I knew by telling my memoirs, I could better educate the public. My personalize experience spanned many in different jobs. The science of water and sustainability is not difficult, but when exploring real stories that explore the everyday life of people just trying to make a living and live a quality life, then the science is much more complex. Reflection is important, especially more so in the discipline of leadership studies. It is through reflection that we begin to see different outcomes and how our own actions contribute to those outcomes. Today, we look to the researcher to awaked and inspire to make contact with and respect their own questions and problems, to determine tactical applications for knowledge and understanding (Moustakas, 1990).

The finish work of autoethnography should read as an intelligent, personal, emotional story that is, supported by previously held knowledge (Sparks, 1996). This is enhancing with adequate justification for the finish work. This is the sharing of unique knowledge that is useful and applicable to a broader context. I hope that the readers of this book processes and connects to the experiences in this book.

We live in a society where false messages about science take over research and discover. Politician and the media rarely explore the truth, but live in their ideology without thinking of the harm this can do. They harm not just to our society, but the future of our children. We spew hate and condemnation quickly! My message is to listen to the scientist, go to the research universities web sites where scientist have been studying water, climate and sustainability for many years. Seek the truth!

I encourage all to experience nature, everyday go to a park, on the weekends have a picnic, watch the snow fall without sound. Kayak down a river, walk in the rain and snow. See the beauty of the forest. To really listen, seek pure knowledge and follow a path of kindness so that the end is noble and good for the greater good. Eckert Tole says. "for the end to meet the definition of the greater good, the path to the end cannot be corrupted".

REFERENCES

Ellis, C., Adams, T. E., & Boucher, A. P. (2011). *Social Research*. Qualitive Review. https://www.qualitative-research.net/index.php/fqs/article/view/1589/3095

Moustakas, C. (1990). *Heuristic research: Design, methodology, and applications*. Sage. doi:10.4135/9781412995641

Sparkes, A. C. (1996). The fatal flaw: A narrative of the fragile body-self. *Qualitative Inquiry*, *2*(4), 463–494. doi:10.1177/107780049600200405

Wall, S. (2006). *An Autoethnography on Learning About Autoethnography*. Sage Journals. https://journals.sagepub.com/doi/10.1177/160940690600500205

Chapter 1
History of Water Science in the USA

ABSTRACT

Water connects all living things. The beautiful natural balance of water, biodiversity, and ecosystems created a beautiful water planet. On this water planet, there is only three percent fresh water, and humans can use less than one percent. Water is the most important resource we have, and one whose sustainability will determine the survival of our species. Native Americans in bygone years knew how to manage our environment. They were connected to the Earth. They knew how to protect the forest from immense fires, they knew how to keep the soils enriched with crop rotation, and they had a connection to nature that all people need to find and develop. The quality of our future could be directly connected to our knowledge and connection to the Earth. By learning our history, we can learn from our mistakes. Through autoethnography, this chapter explores the history of water science in the USA.

"A drop of water, if it could write out its own history, would explain the universe to us". – Lucy Larcome

DOI: 10.4018/978-1-7998-9691-3.ch001

THE CLEAN WATER ACT

In 1948 the Federal Water Pollution Control Act was the first major U.S. law to address water pollution. Growing public concern led to the Clean Water Act being passed in 1972. This act was to address the crisis in national water quality, to restore and maintain the biodiversity, chemical, physical, and integrity of our great waterways. In 1972 amendments were made to the law of 1948. This included: (EPA, 2020).

- Establishment of basic structures for regulating pollutant discharges into the waters of the United States.
- Gave EPA the authority to implement pollution control programs such as setting wastewater standards for the industry.
- Maintain existing requirements to set water quality standards for all contaminants in surface waters.
- Made it unlawful for any person to discharge any pollutant from a point source into navigable waters unless a permit was obtained under its provisions.
- Funded the construction of sewage treatment plants under the construction grants program.
- Recognized the need for planning to address the critical problems posed by nonpoint source pollution.

Revisions in 1981 streamlined the municipal construction grants process, improving the capabilities of treatment plants built under the program. Changes in 1987 phased out the construction grants program, replacing it with the State Water Pollution Control Revolving Fund, more commonly known as the Clean Water State Revolving Fund. This new funding strategy addressed water quality needs by building on EPA-state partnerships. Over the years, many other laws have changed parts of the Clean Water Act. Title I of the Great Lakes Critical Programs Act of 1990, for example, put into place parts of the Great Lakes Water Quality Agreement of 1978, signed by the U.S. and Canada, where the two nations agreed to reduce certain toxic pollutants in the Great Lakes. That law required EPA to establish water quality criteria for the Great Lakes addressing 29 toxic pollutants with maximum levels that are safe for humans, wildlife, and aquatic life. It also required EPA to help the States implement the criteria on a specific schedule (EPA, 2020).

The goals of the law were to eliminate the discharge of pollutants into the waterways, all the nation's streams, rivers, and lakes were to be environmentally sound. We have made headway, but not nearly enough. What is keeping us from going forward? In her groundbreaking book, "Water A Natural History", Alice Outwater describes the history of water and the environment in the United States. She points out what needs to happen, and highlights the great changes made thus far. In her role as an environmental scientist cleaning up Boston Harbor, she learned that industries nationwide were no longer discharging dangerous waste. They had changed their practices and were adhering to environmental laws (Outwater, 1996). However, dirty water is a concern: with municipal sludge, dredging, channeling, and by altering the environment we are still producing dirty water.

Water is the life- source of all living things on our planet. By honoring our environment and realizing the importance of our great resources we not only restore our natural world, but we also create a sustainable planet and a quality of life we so desperately need. Many people do not have pure clean water coming out of their faucets. They look for other sources for their daily water consumption needs. In addition, natural disasters temporarily alter water systems and make them undrinkable. As we look for healthy on-the-go alternatives, many have turned to bottled water. As naysayers protest, others who depend on this source for clean water are grateful for an alternative source when their own tap water cannot be trusted.

THE AGE OF MAN

The rise of mammals is consistent with the extinction of the dinosaurs around sixty-five million years ago. Until then, mammals had a low profile, for around two hundred million years ago. They lived in existence with the dinosaurs. All living mammals, including man, descended from the one line of their species that survived. Mammals did need the extinction of the dinosaurs to diversify into many forms. Man's behaviors and large brains made them adaptable. Man's warm blood and long periods of parental nurturing made him highly adaptable to a wide range of diets (Berkeley, 2020).

The Age of Man came about in what is referred to as the Holocene Period, which is the last ten thousand years of Earth's history, the time since the end of the last major glacial epoch or "ice age". The Holocene has been a relatively warm period in between the ice ages. Another newer name for the Holocene Period is the Anthropocene, the "Age of Man". However, this is misleading because Homo Sapiens are mammals that evolved and dispersed all over the world well before the Holocene Period. Moreover, the Holocene Period

Figure 1. Frozen grain crop
(Getty Images,2000).

has been a part of all humanities history, the rise and fall of all civilizations and the development of human knowledge and technology (Berkeley, 2020). In addition, geologic time is measure by the difference in strata, and some scholars feel this has not been significant enough to go into another era.

LITTLE ICE AGE

There have been small-scale climate and weather shifts, one being the "Little Ice Age" between 1200 and 1800 A.D. A series of massive volcanic eruptions, and changes in Artic has been determined to have created the Little Ice Age. These massive eruptions put particulates into the atmosphere for years, causing what many deemed as nuclear winters. This kept the Earth cool for centuries. Global temperatures fell less than 1 degree C, with parts of Europe cooling more than elsewhere. Growing seasons were reduced by half and winters were brutal. This event caused crop failures and widespread famine throughout Europe. The cooling temperatures disrupted the economic and social structures of the time. Grain harvests did not return to their previous levels for a hundred and eighty years.

This changed centuries of feudal lines, that were largely organized in Europe, to more of a barter, buy, and sell society, bringing in the modern world as we know it. The peasants left estates for food and shelter. Peasants stayed warm by sleeping together and close to their animals. Those who were wealthier used furs as covers and put on cloaks to stay warm in winters. This led to a massive use of resources (Blom, 2017).

In many other parts of the world, the effects of the little ice age were felt. The Ming dynasty fell in 1644, the French Revolution occurred, and President Washington walked across the Delaware River on ice flows. Recently the Delaware River has iced over, but the cause was an Artic blast event. During the Little Ice Age, it was said that birds iced over and fell from the sky. People were dying of hypothermia. In 1588 a hurricane destroyed the Spanish Armada (Blom, 2107) which was contributed to the unrest. At the time, many felt it was the end of earth. The famous and most admired violins, the Stradivarius and Guarneri, were made during this time. Many have determined the trees took longer to mature in the cold, resulting in a denser wood with better sound qualities.

When these empires failed and feudal lines changed it brought on the birth of market and companies, and the socioeconomics changed permanently. These new companies began with the trade of goods, and this started to drive markets. As the market economy grew, the Little Ice Age relinquished it power. As the market economy grew, the natural resources continued to be depleted.

During the little Ice Age furs were worn more than at any time before (Reichholf, 1990). In the Middle Ages, furs had become part of a person's wardrobe, for everyone, even the poor. Moreover, the rich owned many garments. The Vikings might have started this enormous fur trade in the 900's, and it continued to grow and spread out across Europe and the east. In 1390, ninety seven percent of England imported a total of 323,624 furs from March through November. Most of these came from Russia (Outwater, 1996). The beaver was in the highest demand because it was uniquely suited to felting. The undercoat is finer and denser than lambswool or rabbit, and the hair locks together to make the beaver pelt waterproof (Reichholf, 1999). As the little ice age came to an end in Europe, furs were still highly valued, due to the fact that before and during this time fireplaces did not efficiently warm houses or castles. Furthermore, beaver felt hats were high fashion. As the environment became more depleted in Europe, the new world offered plenty of resources.

BEAVERS

Beavers lived throughout the European wilderness and the British Isles. By the late 1300's they were almost non-existent. It was not until better chimneys,

Figure 2. Beaver hat and box
(UK National Archives, 1990)

better glass for windows, and coal heating did the demand for beavers subside, but not the beaver hats. They were still in demand. It was thought that by the mid 1500's the beaver population in England was down to a dozen, even though Russia and Scandinavia still had an abundant population of beaver (Outwater, 1996).

The New World represented a new supply of beavers and other wild animals with rich pelts. In the beginning native peoples were ready to supply furs in exchange for the tremendous amount of new goods that were far superior to their own. Things such as iron tools, cooking utensils, hoes, blankets, axes, gun powder, shot, and glass beads. Unfortunately, beavers were the first to be depleted. The market for beaver hats was strong. Pilgrims of Cape Cod in 1620 were surprised to find that the Native Peoples were stocked with European kettles and other goods (Outwater, 1996). And Native People worked with the fur traders because they laid no claims to the land, they took native wives, and built alliances. There was only one problem, that beavers were not an endless supply. By the eighteenth century the beaver was almost gone in America.

The beaver had almost been eliminated by the colonies in the East, the Hudson's Bay men in the North, and the French with the Spanish in the South (Reichholf, 1999). When they ran out of beaver, they started in on minks, deer, moose, and buffalo. Beavers are great pets, friendly, good natured and clean. They are forty inches long and over a foot tall and have up to five babies at a

Figure 3. Beaver dam diagram
(Science Digest, 2000).

time. Beavers do more to shape the landscape than any other mammal except for man. If you have ever seen an area active with beavers, it is awe inspiring and unnerving. While building their colony, it looks like devastation, but in the end, there is a beautiful wetland. Beavers shape waterways and landscapes, as they build wetlands. These give rise to many organisms and other animals. Wetlands protect against floods, keep shorelines from eroding and cleans waterways. They make meadows out of forests; their wetlands slowly capture silt. They provide food and homes for dozens of species: ducks, frogs, fish, moose, birds and much more. The beaver's ancestors were building dams for millions of years (Outwater, 1996).

When beavers make a series of dams and ponds within a drainage basin, the water cycle in the entire watershed is brought to life. Wetlands are the sponge that soaks up water during storms and releases the water in drier times. These wetlands hold millions of gallons of water. Wetlands clarify the water and prevent the soil from washing downstream (Maxwell, 1994). When spring runoff from melting snow and storms reaches the wetlands, wetlands absorb the runoff and prevent flooding and erosion downstream. Swift water carries silts. Water held in wetlands percolates into the ground and becomes groundwater. With the trapping of beavers, the first significant change occurred to the water cycle in the United States. Today, there are between seven to twelve million beavers in the United States It is estimated that as many as two hundred million beavers once lived in the United States (Outwater, 1996).

Deforestation

Old growth forests began to fall to farming and timber trade. It is said that the amount of Red Wood and Sequoia forests removed, would add 32% of

Figure 4. Beaver dam
(Science Digest, 2000)

rainwater back into the ground water system of California (Outwater, 1996). When water moves through creek beds, streams, and by rivers it supplies the nutrients for forest and organisms. It meanders across the land breaking down matter, to supply other ecosystems which creates harmony between the forest, land, and waterways.

The forest is one of the most important parts to the water cycle. Trees capture water from the atmosphere, it flows through them into the groundwater. In addition, trees capture carbon dioxide in the atmosphere and releases oxygen. As leaves of the tree capture the rain or water vapor from the atmosphere, water is transferred through the tree into the groundwater. Trees protect the soil. Without trees, streams carry less organic matter and more silt. The silt clouds the water, and reduces photosynthesis, and infiltrates the gravel beds, which are the nurseries for fish (Outwater, 1996). As rainwater falls on a forested watershed, the canopy breaks the force of the water and keeps the soil from splashing. Over half of the rain goes directly back to the atmosphere through evaporation and transpiration, creating a microclimate. Some of the rain becomes groundwater and reemerges as spring water, streams that flow into rivers, and eventually the sea. Beneath the forest the soil is honeycombed with channels made not by roots but burrowing animals, earthworms, insects, and fungi. Between 50 to 60 percent of the total volume of healthy forest soil is in this space. Rain trickles down into the space to become groundwater (USGS, 2020).

U.S. Forest Service estimates that about half of the contiguous United States was once old growth forests. Most being primeval trees growing in the Eastern third of our country. Old-growth forest have fallen logs and snags. An old-growth forest has a multilayered forest canopy, with trees that are young to very old. I have seen old growth trees in Maine with the mark of an English king on them. The English ship captains marked trees that would be excellent for building ships, and a few of those trees are still around today, and are protected by the Maine Forest Department. In addition, old growth forests still exist in our great northwest.

In North America, fire was an important part of the natural forest management cycle. Native Americans knew this, and torched large tracts around their villages. Clearing land for crops, stimulating a successional bloom of berries, brows for game, and most important fertilizing the soil. By burning the underbrush, the soil is fertilized and allows for nutrients to be recycled through the soil. Periodic burns allow grasses, shrubs, and non-woody plants to flourish. In addition, it improves conditions for strawberries, blackberries, raspberries, and other gatherable food. Native Americans fired cereal grasses annually, basket grasses and nuts every three years, brush every ten years, large timber on a cycle of fifteen to thirty years or even longer (Pyne, 1982).

Native Americans knew how to maximize the available resources and sustain them for the future. The fires at the edge of forests, increased the meadows. The meadows produced groves of species and a diversity of forest dwellers. These fires created sections that welcomed a host of animal life. Not only did the Native Americans maximize the available resources but they moved seasonally to make them sustainable long-term (Pyne, 1982).

While Native Americans adhered to the natural cycle, the colonist did not understand it. They had a different standard of living. They were growing crops, they cleared forest for villages, and pigs and cattle were turned lose in the forest to forage. As the number of colonists increased so did their hogs. The hogs altered habitat that supplied the Native Americans. You could say the colonist's hogs were the first invasive species in America (Cronon, 1983).

The settlers in New England cleared the forest for timber and fuel. In England, they decimated their forests by the mid 1600's so they needed and wanted the wood from the new world. The sugar plantations in the Caribbean paid a hefty price for wood. By the time of the Revolutionary War, New England had changed from a climax of forest harvested by many Native Americans to an agricultural society that annually exported millions of board feet of lumber (Reynolds & Pierson, 1942).

After 1815, the vast forest between the Mississippi Valley and the Appalachians was sold off, logging commenced, and most of the land was converted to farms. Some states were almost completely deforested. In the meantime, a new movement was being born, *Transcendentalism.* This belief presented the wilderness and forests as essential to creation and our quality of life. The famous writers Ralph Waldo Emerson and Henry David Thoreau were leaders in this movement. They, together with the painter James Fenimore Cooper and other writers and painters depicted the wilderness as a place of solitude, intellectual pursuits, and spiritual regeneration. John Muir's writings in the late 1800's gave rise to the protection of our great wilderness areas. It was President Theodore Roosevelt who created the first national park (Outwater, 1996).

Old growth forests remain in only a few areas of the United States. Many states have replanted and protected their forest; some have come back to be over ninety percent forested. The Northeast, Maine, New Hampshire, and Vermont have established forestry departments and NGO's who actively burn and manage diverse forests. The states that do not actively burn, have not fared well. Raging wildfires have been harder to contain. By not burning the slash on the forest floor and not doing prescribed burns, the fires are hotter and more prolific (U.S. Gov, 2020). The most significant change by deforestation in the United States may be the waterways, that no longer feed many trees on their way to the sea. The productivity of forest streams may be disappearing (U.S. Gov, 2020).

TALLGRASS PRAIRIES

As settlers moved west, they encountered the tallgrass and shortgrass prairies. Scandinavian settlers going to Kansas were amazed by the prairie grass, it was taller than their covered wagons. They wrote about the violent lightening and storms across the great plains. Lighting was an important part of the ecosystem.

Spring storms produced lightening which would cause extensive burning, which in turn fertilized the plants at the start of the growing season. Humans and lightening have lived in co-existence for thousands of years. Fire, caused by lightening, encouraged the growth of grasses instead of woody plants, shrubs, and trees. A tall grass prairie will revert to scrub woodland in a matter of decades without regular fires. To further sustain the grasslands the Native Americans had a deliberate routine of setting fires. These fires were used for many reasons: hunting, warfare, agriculture, and to modify habitat. These grasslands extended from the Mississippi River to the Rocky Mountains (Filson,1966).

The barrens of Kentucky, the eastern outpost of the tallgrass prairies, was a crescent-shaped meadow of 6,000 square miles surrounded by virgin

Figure 5. Deforestation
(Adobe Stock, 2021).

forest. It was burned annually by the Native Americans. The land supported thousands of buffalo. When the settlers moved in, the buffalo herds were annihilated, and the land treed to forest. The plains original vegetation varied according to place, but consisted of tall grasses, short grasses, tall forbs, and

Figure 6. Tall grass prairie
(National Parks, 2000).

short forbs. These plants live together in an ecological system, that is now gone (Outwater, 1997).

These prairies fed large groups of buffalo, deer, and antelope. To protect themselves from insect bites, the buffalo would dig shallow holes around 20w -x2d (wallow). The holes would collect rainwater. What did not evaporate would go into the groundwater system. The wallows were dug in low areas, and they collected runoff as well as rainwater. The water that pooled in the wallows was very different from the rain that fell on the great prairies. Only the uppermost layer of water in the wallow would evaporate, the rest would seep down into the water table. This created a zone of saturated soil that extended from the bottom of the wallow to the groundwater. Basically, every wallow was a recharge pond. Many geologist and engineers have tried to replicate the wallows, but they filled with silt. However, the buffalo's hooves prevented a silt layer, forming a perfect groundwater recharge pond (Filson, 1966).

Most of the rain falls on the Great Plains in the spring and summer. Most of the rainfall is caught by vegetation, where it either evaporates or drips down into the soil. Some of the water will runoff as surface water. Because of the climate, rainfall rarely percolates below the roots zone to replenish the groundwater. Because of the lack of trees, and regular fires, many smaller animals lived below ground (Outwater, 1997).

PRAIRIE DOGS

Prairie Dogs are like the beavers, they significantly alter ecosystems and provide habitat for other species. It is thought that at one time billions of prairie dogs inhabited the Great Plains. Prairie dog towns extended for thousands of miles, with up to fifty holes per acre. These towns are underground mazes of five-inch diameter tunnels usually lying in the root zone, which may reach ten to twelve feet deep. In the process of constructing their maze of tunnels, the prairie dogs once moved tons of topsoil and organic matter, including clipped grass, roots, feces, urine, insect parts, and other byproducts of life. They made loam in a sandy dry world. The greens between holes were kept clipped to expose predators, creating a short-grass plant community. They provided new growth and protein for the surrounding grasses, so buffalo and cattle grazed on the byproducts of prairie dog towns (Whicker & Detling, 1983).

When water falls on a prairie dog town, if falls on soil filled with macropores or tunnels larger than a millimeter in diameter. Elsewhere, the entire soil

Figure 7. Praire dogs
(National Parks, 2000).

profile, from the surface of the earth down to the water table, needs to be saturated before water can percolate down to the table freely, as it does in a buffalo wallow. Macropores naturally promote rapid transport of water through the soil. High-intensity short duration rainfalls in the Great Plains enter macropores and rapidly funneled below to the root zone. The prairie dog population increased the amount of rainfall percolating down to the groundwater and feeding the region's streams and rivers.

The great grasslands are a complex system. Americans of European decent had never seen the great buffalo herds. These great herds were shot for their tongues and the rest of the meat was left to rot. Hides were sent east, and soon the great herds were gone. Cattle became the mainstay of the great plains, followed by wheat. When the keepers of the ecosystem leave then changes to the ecosystem are not good. The water that flowed clear when the buffalo herds roamed, became muddy and rank. Fish no longer could exist as temperatures rose. Silt covered the gravel beds where fish had spawned, and other wildlife started to disappear. Prairie dogs were a nuisance and were poisoned. The rate of groundwater recharge dropped, there was less groundwater to feed streams (Whicker & Detling, 1983).

In the mid 1800's thousands of immigrants moved to the Great Plains from countries due to persecution and poverty. These settlers brought hard

Figure 8.
(History, 2020).

red winter wheat, which required less moisture. This came to dominate the crop production of the Great Plains. From 1878 to 1887 the rains came, year after year, and farmers moved further to the west. At the time, the settlers moved into the great plain, they were experiencing an average rainfall of twice above normal. This was a weather event, that was followed by a five-year drought. After this came mechanized farming and the Great Plains were plowed up. In the end, poor farming practices and dry conditions led to the Dust Bowl (History, 2020).

Which in turn led to the discovery of the Ogallala aquifer. This aquifer is the remains of a vast alluvial plain that stretched east from the Rockies. As the climate grew warmer and the glaciers started receding, erosion slowed, and the Rio Grande and Pecos Rivers eventually cut through the alluvium faster than it was being deposited. Those deepening rivers diverted the Rockies' runoff, which once had flowed into the Ogallala deposits, cutting the Ogallala off from its recharge area. New irrigation methods were introduced around the time of the Dust Bowl and the Ogallala aquifer was discovered. Today it is almost depleted. It is not renewable, due to its unique geologic structure. Those that depend on it today, will not have it in the future. It is the water source for several states, ranchers, and farmers (History, 2020).

DAMS

The redirections of water resources when the great dams were built, and water was re-directed to feed large cities and produce hydropower, changed our waterways indefinitely. Many civilizations tried great projects where water was diverted for irrigation. The great civilizations of Mesopotamia and Anasazi, both failed long term in the re-direction of water for irrigation, it was not sustainable (Adams, 1965).

Mesopotamia means "between the rivers". The Mesopotamian civilization developed between the Tigris and Euphrates rivers, and this area was known as the Fertile Crescent, which was a large rich area of farmland extending from the Persian Gulf to the Mediterranean Sea. These farming settlements were established around 7000BC. This development was possible because of the flooding of the Tigris and Euphrates rivers. The flooding left a fertile layer of soil for crops and provided food for early inhabitants. Irrigation techniques helped farmers because it provided a way of supplying water to fields and storing water for periods of time, which made the farmers more productive. This led to a food surplus and less people were needed to farm (MSU, 2014).

In ancient Egypt, the Nile River was the main source of water. When the Nile's flow was high, a system of canals and dikes was built to flood large tracts of land. The uncertainty of the Nile's flow led to the canals and dikes being washed away and villages flooded. During low water times, the crops could not grow. In many places the fields were too high to receive water, and no crops could grow. Water was drawn from the canals or the Nile directly (Adams, 1965).

In 900 C.E, the Anasazi of what is now New Mexico developed a large civilization which consisted of more than two thousand four hundred archaeological sites with nine towns each containing hundreds of rooms along a nine-mile stretch. Water was limited, so a collection system ran down from the side canyon that drained from the upper mesa top and was diverted by either an earth or masonry dam near the mouth of the side canyon into a canal system. These canals averaged about fifteen feet wide and more than four feet deep. They were lined with stone slabs and bordered in other areas by masonry walls. Water was then diverted to the fields in small ditches or into overflow ponds and small reservoirs (Doolittle, 1990). Little is known about this civilization, and why it disappeared. A long-term draught would have been detrimental. Draught comes to this area frequently; it is a common weather event, or they just depleted a system called the Chacoan water system,

In 1902 the Reclamation Act was created to determine how to irrigate the unfarmable deserts of the West. In 1903, six major irrigation projects were approved. By 1919 there were twenty-six similar projects approved. The concept of reclaiming a desert was a task given to the best engineering graduates of the time. The cost and size of the projects increased, and by 1991 a total of 339 reservoirs, 154 diversion dams, 7,670 miles of irrigation canals, 1,170 miles of pipeline, 270 miles of tunnels, 267 pumping plants, and 52 hydroelectric power plants were built. Water made much of the land prosperous and productive (Outwater,1996). But at what cost?

When a watershed is dammed and lakes and reservoirs fed by a river replace the swift turbulent, highly oxygenated water of a free river, the species that call these rivers home will fade away. Soon, the entire aquatic community of a dammed river will begin to resemble that of a lake. When rivers are dammed the fish that swim upriver to spawn will begin to disappear. Fish ladders have been built to allow the fish to swim upriver, but they have never been successful enough to bring back the many species that have disappeared. When the fish disappear, so do many other species that are sustained by the fish. Dams and altering our waterways were one the most serious catastrophes regarding the sustainability of our natural resources (Blate, 2020). Today, dams are being removed from several states. In 2020 sixty-nine dams were removed to bring back the natural habitat and ecosystem. Many of these dams were outdated and not well maintained. The Mississippi that once had large thriving fish populations has been transformed to a river that hardly moves due to damming, channeling, and levees.

MAN-MADE SYSTEMS

Problems with sewage and waste have been around for thousands of years. Water supply systems and sewer systems allowed cities to clean streets by washing the waste into bodies of water that were close. These water systems created a new way of living, flush toilets created great hotels and destinations. Modern bathrooms use hundreds more gallons of water each day. More trash and waste were dumped into rivers. Eventually, modern pesticides and fertilizers found their way into rivers which inevitably goes into the oceans.

In 1962 Rachel Carson's book Silent Spring exposed the devastation to our ecosystems and environment. This initiated the environmental movement. It led to the 1965 Water Quality Act, which in 1972 led to the Federal Water Pollution Control Act called the Clean Water Act. Today, every city in the

U.S. is required to build and operate a wastewater treatment plant. In addition, provide safe drinking water and air quality (U.S. Gov, 2020).

MAN-MADE PROBLEMS

One third of the freshwater mussels are found in the U.S. waterways. Freshwater mussels make pearls and line their shells with a beautiful array of colors. The First Peoples of North America gathered the mussels for food, pearls, and wampum. At one time the U.S. waterways were abundant with mussels. In 1857, a carpenter found a large pink pearl in Notch Brook, near Paterson, New Jersey. It soon became known as the Queen Pearl and was purchased by Tiffany & Company. In 1870 an anthropology professor found a large cache of freshwater pearls, which led to many farmers looking for pearls.

The mighty Mississippi was once home to the finest pearl producing mussels. In 1891, Johannes Boepple started the first shell-button industry with a small factory on the banks of the Mississippi River. The business grew, factories sprung up, and beds rapidly became depleted. By 1897, within three years the bed used for the buttons was deplete, 167 miles of river mussel beds were depleted (Outwater, 1996). In 1907, the federal government funded a program to artificially propagate the rivers with mussels. By 1930, the Bureau of Fisheries abandoned it's mussel-stocking program, due to its failure. In the end, the over harvesting of the mussel beds had left them on the endangered species list (Outwater, 1996). The problem with this is that mussels clean our waterways. An invasive species might have taken their place.

There is one mussel that has thrived, the zebra mussel. Its origin starts in the Caspian Sea. It spread out through European waterways and eventually made its way to Lake Erie and Lake Huron. It most likely attached itself to ships traveling from Europe to the U.S. Unlike the freshwater mussels native to the U.S. which bury themselves in the mud of rivers, the zebra mussels cling to surface water by attaching to ship hulls or intake pipes. While native mussels are disappearing, the zebra mussels are reproducing in great numbers. Each mussel filters about a quart of water a day. The zebra mussels are starting to clean the rivers. The success of the zebra mussels will only mean the elimination of native mussels.

Today we have many more invasive species. We have flying fish in the great lakes that are killing off native fish species. Birds, plants, and animals are being illegally introduced yearly. Now we have agencies working tirelessly to combat these issues. Each state has several departments leading the charge

Figure 9. Peal buttons
(Ohio Historical Society, 2022).

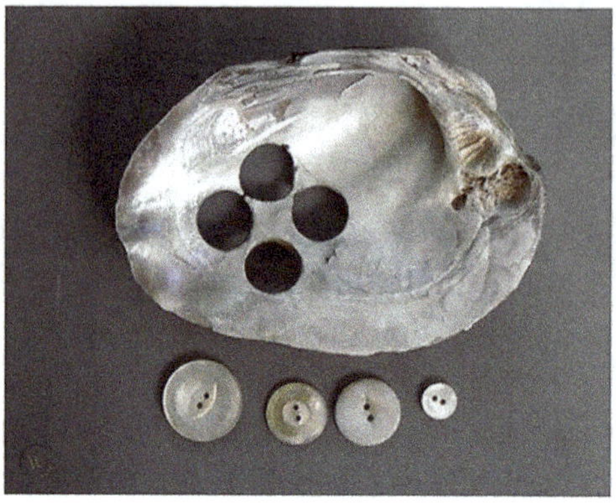

to protect our native habitats and ecosystems. There are federal agencies such as the Forest Service, Fish and Wildlife Service Bureau of Land Management and the National Park Service. What it all comes down to – people will continue to do wrong, unless they are educated to the right way of managing our ecosystems. They must know and understand how nature is connected.

REFERENCES

Adams, R. M. (1965). *Heartland of Cities, Surveys of Ancient Settlement and Land Use on the Central Floodplain of the Euphrates*. University of Chicago Press.

Berkeley. (2020). *Hall of the Mammals*. University Campus Berkley. Retrieved from https://ucmp.berkeley.edu/mammal/mammal.html

Black, R. (2012). *Volcanic Origin for Little Ice Age*. BBC. Retrieved from https://www.bbc.com/news/science-environment-16797075#:~:text=The%20Little%20Ice%20Age%20was,Arctic%20ice%20cover%2C%20scientists%20conclude.&text=They%20say%20a%20series%20of,for%20ice%20sheets%20to%20exp

Blate, J. T. (2021). *69 Dams Removed in 2020*. Americans Rivers. Retrieved from https://www.americanrivers.org/2021/02/69-dams-removed-in-2020/

Blom, P. (2017). *Natures Mutiny, How the Little Ice Age of the 17th Century Transformed the West and Shaped the Present*. Carl Hanser Verlag Muchen.

Cronon, W. (1983). *Changes in the Land: Indians, Colonist, and the Ecology of New England.* Farrar, Strauss, & Giroux.

Doolittle, W. E. (1990). *Canal Irrigation in Prehistoric Mexico*. University of Texas Press. doi:10.7560/715585

EPA. (2020). *Clean Water Act.* The Environmental Protection Agency. Retrieved from https://www.epa.gov/laws-regulations/history-clean-water-act

Filson, J. (1966). The Amazing Herds of Buffalo. In *The Discovery and Settlement of Kentucky*. Ann Arbor University.

U.S. Forestry. (2020). *National Forest*. U.S. Gov. Retrieved from nationalforests.org/blog/what-are-the-differences-between-national-parks-and-national-forests?gclid=EAIaIQobChMIq6jWr4if7wIVpvLACh109QAiEAAYASAAEgIYn_D_BwE

Gore, R. (2020). The Rise of the Mammals. *National Geographic.* Retrieved from https://www.nationalgeographic.com/science/article/rise-mammals

Maxwell, J. (1994, Mar.). Leave it to the Beavers. *Audubon*.

MSU. (2014). *The Role of Irrigation in Mesopotamia.* MSU. Retrieved from http://anthropology.msu.edu/anp363-ss14/2014/02/27/the-role-of-irrigation-in-mesopotamia/

Outwater, A. (1996). *Water A Natural History.* Perseus Book.

Pyne, S. J. (1980). Firs in America: A Cultural History of Wildland and Rural Fire. Princeton University Press.

Reichholf, J. (1990). Beavers. In Grzimek's Encyclopedia of Mammals (vol. 3). McGraw Hill.

Reynolds & Pierson. (1942, February). ATimber Use in the Eighteenth and early Nineth Centuries. *USDA Circulation*, 641.

Whicker & Detling. (1983). Advantages of Grazing Cattle in Prairie Dog Towns. *Wildlife Management*, *47*, 620–645.

Chapter 2
Water Right Law

ABSTRACT

Water is the most important resource on our planet. We are considered the blue planet. Only three percent of our water is freshwater. Of the three percent, only one percent is accessible for human use. Water makes up 45-75% of our body. Seventy percent of our brains are water, with 80-95% of our blood being water. Clean pure water is important to our overall health and quality of life. Water is the heart of agriculture. The availability of freshwater makes it possible to grow crops and raise livestock. Water use, in turn, is at the heart of heated discussions involving water law and policy. Water is for all users, and even though is it plentiful, there is often not the right quantity or quality in the right place at the right time to meet demand. Moreover, competition among water users, including agriculture, municipalities, industry, recreational users, and conservationists can become volatile. Through autoethnography, this chapter explores water rights laws.

"All the water that will ever be is, right now" – National Geographic, October 1993

Water is the most important resource on our planet. We are considered the blue planet. Only three percent of our water is freshwater. Of the three percent, only one percent is accessible for human use. Forty-five to seventy-five percent of our body is water. Seventy percent of our brain is water, with eighty to

DOI: 10.4018/978-1-7998-9691-3.ch002

Figure 1.
http://www.sustainablewaters.com/the-water-footprint-walking-lightly-to-save-the-earth-and-your-money, 2011).

ninety-five percent of our blood being water. Clean pure water is important to our overall health and quality of life. Traditionally, the management of water resources has focused on surface water or groundwater as separate ecosystems. Water tends to be regulated based upon the source, depending on the category of perennial, ephemeral, or man-made. More broadly, water law can generally be divided into two substantive areas: rights to use water and restrictions on pollution of water. Water concerns range from a balance in public and private rights to water use. The right of an individual land owner to rights of owners downstream. As it relates to agriculture, water law tends to fall into two categories: allocation rights and agricultural land use that could negatively affect water quality.

AGRICULTURE

Water is the heart of agriculture. The availability of freshwater makes it possible to grow crops and raise livestock. Water use, in turn, is at the heart of heated discussions involving water law and policy. Water is for all users, and even though it is it plentiful, there is often not the right quantity or quality in the right place at the right time to meet demand. Competition among water users, including agriculture, municipalities, industry, recreational users, and conservationists can become volatile.

Agriculture is a major user of ground and surface water in the United States and globally. In the U.S. alone it is estimated that eighty percent of the nation's consumption of water use is in agriculture, even more for the western half of the U.S. where over ninety percent of water is used in agriculture. Water use is sourced from rivers, lakes, streams, and ponds. In addition, agricultural water is sourced from rainfall and stored in cisterns.

Water allocation is generally governed by the states, with each state having its own regulatory system with very little federal intervention. State statutes and regulatory schemes control certain uses of water, such as transfers of water from one watershed to another, and withdrawal of groundwater from overused aquifers, cisterns, and wells. However, water quality is regulated by the federal Clean Water Act.

SURFACE WATER RIGHTS

There are three different allocation systems to determine the rights of private individuals with regard to water use. The first is the riparian doctrine, which started in the water-rich eastern part of the United States. The second is the system of prior appropriation or "first-in-time, first-in-right", which started in the western part of the United States where water is not so plentiful. In addition, a handful of states have adopted a hybrid system, which contains parts of both. Water allocation is complex, and for this reason, it is the states that have developed and revised the legal, policy, and regulatory rights of competing water users (Water Education Foundation, 2021).

THE RIPARIAN DOCTRINE

The riparian doctrine limits the usage of water to only those landowners with riparian land. To be classified as a riparian landowner, the land owner must own the land adjacent to the watercourse (river, stream, lake, or pond), from which the landowner plans to use the water. However, the water may only be classified as of reasonable use. The courts can settle unreasonable uses between landowners. A landowner cannot interfere with the reasonable use of another riparian landowner. Reasonableness is determined by comparing the proposed use with the other uses of other riparian landowners. Natural uses, such as drinking water, livestock use or watering a garden are considered reasonable uses under the law. Irrigation or industry are considerable reasonable uses under most state's law.

Riparian rights gained legal prominence after California became a state. Recognition of riparian rights encouraged agriculture development in California's fertile river valleys, especially after the 1862 Homestead Act, which transferred large amounts of land from public to private ownership. In the 1880's, two of California's largest water users went to court over water rights in the Tulare Basin. California upheld the riparian rights. Riparian rights are considered "part and parcel" of the land and are passed with the title to the property. This limits the water from being applied to the most economical use (Water Education Foundation, 2021).

Courts have permitted riparian rights owners to agree not to use water in which they are entitled, so that others may claim the water under another right. The riparian right holder usually is compensated in a manner very similar to a sale of the water right. This system was established in 1914. At the time, the Water Commission Act formalized the appropriation system and centralize appropriative water right state records. Under the act, the state required new appropriators to obtain a permit from the state prior to diverting water. Since the riparian rights are not lost to the non-use owner, the owner has the right to use the water any time he deems necessary (Water Education Foundation, 2021).

Non-riparian rights, landowners that have no right to use water, although some riparian jurisdictions might allow usage. Many jurisdictions require proof of harm from the use of water on non-riparian land. This follows the Restatement (Second) Torts # 855 which allows for the reasonable use of water on non-riparian land only if the user also owns riparian land. Under riparian rights, landowners do not have to use water by obtaining a permit. A central state agency controls who may use the water, how much they can use, and when they can use it. This is called a regulated riparian system. State regulatory agencies will determine if the use is reasonable by taking in account both the potential benefits to society and the compatibility with current uses before granting a new permit. In most cases, the permit is only required on consumptive uses and excludes non-consumptive uses, or uses that do not require a diversion or removal of water from the watercourse (National Center for Agriculture, 2021).

This permitting system allows for future planning and best practices of the watercourse. These rules and granting of permits vary from state to state. These regulated riparian permits are for a fixed period of years. In times of drought, adjustments on the quantity of water withdrawal will be made to prevent harm. If this is mandated and the permit holder does not comply, they may have to forfeit their permit.

THE PRIOR APPROPRIATION DOCTRINE

The prior appropriation dates back to miners who first settled the western United States, and needed water to develop their mining claims. Due to land not being adjacent to a watercourse, the miners could not use the riparian system. The "first-intime, first-in right" system was developed. This led to many disputes over water uses. This led to the prior appropriation doctrine, where the first user had the right to continue using the water to the exclusion of the rights of those who came later. The prior appropriation system is based on priority. The senior appropriator has the highest priority and can defeat all other less senior appropriations in times of shortages. There is no requirement that a senior appropriator use less water in times of shortages. Users can take, in order of their respective priorities, with each taking their full appropriative right until the water is gone. The theory being that it is better for water to be used by the junior appropriator rather than lost in transport to the senior appropriator.

The prior appropriation doctrine is determined by states and may vary from state to state. A beneficial use of water is the measure and limitation of the appropriate right. Once water is put to a beneficial use, the right is perfected and has priority over later appropriators. However, the senior appropriator then has the right to use their original right, even if a better use arises (Water Education Foundation, 2021).

HYBRID SYSTEMS

States such as California and Oklahoma, have developed hybrid allocation systems. These combine aspects of both the riparian and the appropriative rights systems. These are not uniform; they just contain elements of the other systems.

PRIVATE GROUNDWATER RIGHTS

Groundwater aquifers are connected to springs and surface water systems. There are multiple rules and policies used by states to allocate and regulate groundwater. These include the Absolute Dominion rule, which is Correlative Rights doctrine, Prior Appropriation doctrine, Reasonable Use doctrine, and

Restatement of Torts rule. Groundwater allocation systems often differentiate between on-tract uses. On-tract use is where water is used on the tract where the pump is located. Off-tract use is where water is transferred to another location for use. Many states may not fall clearly within a particular doctrine, and may use components of two or more systems (National Center for Agriculture, 2021).

ABSOLUTE DOMINION RULE

This is also called the "Absolute Ownership Rule" or the "English Rule". A landowner may use as much groundwater as possible. The rule does not take into account impacts on neighboring uses, and as a result, one owner could monopolize the entire aquifer incurring liability. This doctrine creates an incentive to pump as much water as possible because of the lack of concern of incurring penalties from a neighboring user. Most states have rejected this doctrine, seeing that this could bring about harm (National Center for Agriculture, 2021).

CORRELATIVE RIGHTS DOCTRINE

This doctrine distributes water on an equitable basis among landowners and allows off-tract uses. These off-tract uses are subordinate to on-tract uses. Like the Absolute Dominion Rue, the Correlative Rights Doctrine determines rights in groundwater based on ownership of land. The landowner overlying the aquifer is limited to a reasonable share of the aquifer's total supply, rather than having an absolute right to unlimited withdrawal of the aquifer (National Center for Agriculture, 2021).

PRIOR APPROPRIATION DOCTRINE

In the western part of the United States, the prior appropriation doctrine has been adopted by many states. The first landowner to beneficially use or divert water from a groundwater source is given priority over later users. The right is limited to reasonable use. States are beginning to replace this doctrine with a permit system, similar to the surface water permit system. This doctrine is in use in Alaska, Colorado, Idaho, Kansas, Montana, Nevada, New Mexico,

North Dakota, Oregon, South Dakota, Utah, Washington, and Wyoming (National Center for Agriculture, 2021).

REASONABLE USE RULE

In some states the Reasonable Use Doctrine has been adopted, which requires the water to be put to a reasonable use on the overlying tract of land and does not permit water to be taken to another tract. Reasonable use means as long as the water is used on the overlying land. The rule is considered a modification of the Absolute Dominion Rule without the exceptions of wasteful uses and off-tract uses. This system is used in Alabama, Arizona, Delaware, Illinois, Kentucky, Maryland, Michigan, Missouri, New Hampshire, New Jersey, New York, North Carolina, Oklahoma, Pennsylvania, Virginia, and West Virginia. In Florida, the common law groundwater rights have been abolished, and now a permit system is in place. Wyoming, uses the reasonable use doctrine along with the Prior Appropriative system for groundwater. Nebraska, uses the reasonable use doctrine along with the Correlative Rights Doctrine (National Center for Agriculture, 2021).

THE RESTATEMENT (SECOND) OF TORTS RULE

This is a merger of the Absolute Dominion Rule and the Reasonable Use Rule. This is a liability for use of groundwater. (National Center for Agriculture, 2021).

- The withdrawal of groundwater unreasonably causes harm to a proprietor of neighboring land through lowering the water table or reducing artesian pressure
- The withdrawal of groundwater exceeds the proprietor's reasonable share of the annual supply or total store of groundwater
- The withdrawal of the groundwater has a direct and substantial effect upon a watercourse or lake and unreasonably causes harm to a person using its water

PRIVATE WATER HARVESTING RIGHTS

Rainwater harvesting involves capturing, diverting, and storing rainwater from rooftops for later use. This includes agricultural irrigation. This practice is not regulated by the federal government and individual state regulations may vary. Many states do not regulate rainwater collection, or when they do, they allow and even encourage it by offering tax credits or exemptions for the purchase of harvesting equipment. States, especially prior-appropriation jurisdictions, place restrictions on the amount of rainwater that can be collected and/or the method used for collection. Due to advances in technology, atmospheric water harvesting (AWH) or atmospheric water generation (AWG) is a potentially viable agricultural irrigation tool. This involves the capture and collection of small airborne water droplets or vapor using sorbents or mechanical refrigeration technology. Fog and dew harvesting are examples. AWH is not federally governed nor specifically regulated by the States at this time; however, as technology efficiencies increase, the tool is likely to garner regulatory attention (National Center for Agriculture, 2021).

PUBLIC RIGHTS TO WATER

Public water rights fall into four categories. Rights associated with navigation, public trust doctrine, reserved water rights, and public interest protection. Navigation servitude is a United States constitutional law doctrine that gives the federal government the right to navigable waterways as an extension of the Commerce Clause. This is designed to keep waterways open for commerce navigation. It creates a dominant property right held by the federal government for the benefit of the general public. The Public Trust Doctrine is a common law doctrine rooted in Roman law to ensure certain natural resources, like navigable waters are preserved for the public benefit. The states act as a trustee of the common resource and have an obligation to manage it for the benefit of current and future generations. Many states have embedded this doctrine into their own constitutions to protect navigable waterways. This being the *public trust* applied to commerce and fishing in navigable water, but in many states, it's uses have been expanded to include recreation (National Center for Agriculture, 2021).

RESERVED WATER RIGHTS

This is the ability to fully develop water resources through irrigation of croplands or other projects. This can be severely limited by federal reserve rights of water. Federal actions reserving public lands create a water right that allows for enough water to accomplish the purpose of the reservation. This would include national parks, monuments, forest, wild and scenic rivers. In addition, this includes Native American reservations. The priority date is the date the reservation is completed. Private rights established prior to the reservation have priority over the reserved rights. The federal right cannot be abandoned or lost through nonuse. Once asserted, it can take water from private right-holders whose rights were established subsequent to the reserved right (National Center for Agriculture, 2021).

Under the McCarran Amendment, 43 U.S.C. § 666, the federal government has waived sovereign immunity for the limited purpose of adjudicating western water rights. This amendment authorizes the United States in comprehensive stream adjudications, in both state courts and state water agencies. This amendment has been construed to include the water rights of Native Americans to reduce the number of court proceedings to determine the same rights. The Amendment allows for the more efficient adjudication of water rights along streams in the Western States.

PUBLIC INTEREST PROTECTIONS

Some states mandate public interest review of initial water right allocation or increases, requiring an appropriation permit from an administrative agency and allowing the permit to be issued only if the proposed appropriation conforms to the public interest or welfare. Historically, public interest criteria were satisfied if the permit applicant would benefit economically from the water use. More recently, public interest criteria are expanding to include environmental and other public concerns, requiring consideration of the cumulative effects of water withdrawals from ground or surface waters. Similar protections include state statutes creating minimum streamflow requirements or authorizing instream flows (National Center for Agriculture, 2021).

WATER POLLUTION

Water pollution is complex and regulation is comprehensive. It involves areas such as common laws of nuisance, trespass, and negligence, but more often it involves an interconnected network of federal and state statutes and regulations. The cornerstone is the Clean Water Act (CWA). The Clean Water Act provides a comprehensive system for the regulation of pollutants in the waters of the United States with the objective of restoring and maintaining the chemical, physical, and biological integrity of the nation's waters. The CWA operates by authorizing water quality standards for surface waters, requiring permits for point source discharges of pollutants into navigable waters, assisting with funding for construction of municipal sewage treatment plants, and planning for control of nonpoint source pollution (EPA, 2021).

Water rights vary from state to state, so much so, that there are differences in how water in managed, permitted and allocated, even within different watersheds in the same state. Some states allow for a water right to be forfeited if they are not used in a certain timeframe, or require a new application. Other states do not. Basically, a water right on paper is not necessarily a water right in practice. When closing a real estate transaction, it is important to understand and have a thorough knowledge of the water rights associated with the property. In addition, knowledge on land use practices are important to avoid polluting water resources (EPA, 2021).

Ann Castle is a water attorney in the state of Colorado, a state in the American West where water is more valuable than gold. Castle is one of seventy full-time water law attorneys in the state of Colorado. She is involved is some wild disputes, but her commitment to civility has severed her well. Her clients range from ski areas, municipalities, churches, mining companies, real estate developers, ditch companies and ranchers. In 2004 she was able to get several Colorado jurisdictions to agree to an unprecedented regional water management plan. This plan included the Southeast Colorado Water Conservation District, Pueblo Board of Water Works, and the towns of Pueblo, Aurora, Fountain and Colorado Springs (Immel, 2004). This kind of collaboration is an example of future practices. It may result in a complicated plan, but plans and collaborations such as this are needed to solve complicated issues. It is far better to collaborate than to fight lengthy court battles.

Many parts of the Western United States have experienced long-term drought. In 2021, Florida, parts of New England and the northern middle states are being impacted by drought. The long-term relentless drought in the West has had a devastating impact to the social economics of the region. Canals that carry water to supply agricultural areas and municipalities throughout the fertile Central Valley and Southern California (State Water Project and Central Valley Project) have not been able to meet the allocations of water. This has resulted in farmers having to fallow more than five hundred thousand acres of prime agricultural land. Many rural towns have come close to running out of water (Fletcher & Fulford, 2017).

Several states have been impacted severely, particularly those in the Colorado River Basin. These are Arizona, California, Colorado, Nevada, New Mexico, Utah and Wyoming. The Colorado River Basin supports approximately forty million people and four million acres of irrigated agriculture. According to the Bureau of Reclamation, Lake Mead and Lake Powell are an example. The historic low levels of both lakes, the lowest levels in modern times can impact agriculture. Both lakes provide irrigation to around twenty seven percent of the United States national GDP (Bureau of Reclamation, 2021).

Water fuels the American economy as much as oil. The west and Florida produce most of the nation's food. California alone produces nearly half of the total U.S. vegetable crop and more than half of the fruit and nuts (California Department of Food and Agriculture, 2014).

Water conservation at the forefront is the states most affected by the long-term drought. Fletcher and Fulford proposed five proposals to encourage the use of market driven strategies to water management: (Fletcher & Fulford, 2017).

1. Reform legal rules that discourage water trading to enable short-term transfers. Western water law creates significant obstacles to water transactions that will take many years to reform. This would allow a more flexible use of water resources by authorizing simple, short-term water transactions.
2. Create basic market institutions to facilitate trading of water. Promote longer-term water transactions and transfers, state and local governments should establish essential market institutions, such as water banks and exchanges.
3. Use risk mitigation strategies to enhance system reliability. Water managers should support and encourage the use of market-driven risk management strategies to address the uncertainties of water supplies.
4. Protect groundwater resources. To support the development of new programs and markets the states need to better regulate the use of groundwater. This can be done by monitoring and limiting use.

5. Continue and expand federal leadership. Strong federal leadership is necessary to promote interstate and interagency cooperation in water management.

Recent court cases in New Mexico about prior appropriation, which is a cornerstone of the state's water law, is just the beginning of states trying to find solutions to water shortages. Two law suits specifically were against the state engineer who is trying to find solutions to the state's complicated water issues. John Antonio, the New Mexico state engineer attempted to regulate water through a set of rules called Active Water Resource Management, stating it was constitutional. The Tri-State Generation and Transmission, the New Mexico Mining Association and the Middle-Rio Grande Conservancy District filed a law suit against the actions of the state engineer. The other law suit was filed by Horace Bounds Jr. against the state engineer. Claiming the state was in error in permitting new domestic wells in an area where water is fully appropriated and limited (Matlock, 2010).

The century old law of prior appropriation, "first-in-time and first-in-right" is behind the law suits and disagreements. Whoever first puts water to a beneficial use, such as irrigating a field gets first dibs on the water in any given year. Senior waters rights stay with the oldest user. New Mexico state pueblos and tribes have the oldest water rights, followed by farmers, then cities, power companies and new developments. The current state engineer says all of New Mexico's water rights are fully appropriated. With populations growing in cities, the farmers rely on the state to enforce their priority water rights (Matlock, 2010).

The Tri-State law suit came from this dilemma. The state legislature passed a law giving the state engineer the authority to create a water-management program with rules, as long as they did not impact the senior water-rights holders and did not increase water depletions. The actions of the state engineer allowed for water to be provided to junior users while still protecting senior water rights. This was done with permitting, surveys and agreements.

A District Court found the state engineer overstepped his authority and violated the process. The state engineer appealed to the State Supreme Court, and the judges found the state engineer can use only adjudications and licenses to determine the priority of water rights, basically making the states Active Water Resource Management program null and void. This limits the state engineer to protect senior water rights (Matlock, 2010).

In the Bounds case, the law requires the state engineer to issue a domestic well permit to any landowner who applies for one without determining if there's any water available or if pumping the well would impact a senior water user. Bounds claimed new wells were depleting surface water on the Upper Mimbres River and impacting his senior water rights. The Tribes joined the argument. A lower court was in favor of Bounds, but the New Mexico Appeals Court disagreed, saying the priority doctrine was a broad principle, and the legislature is authorized to enact laws providing water management. Bounds appealed to the State Supreme Court (Matlock, 2020). He gave up his appeal after the decision by the State Supreme Court (Case Text, 2021). There are currently over one thousand law suits filed in the State of New Mexico over water law rights (Case Text, 2021). This is evidence of the complexity of water rights.

An interstate water compact is an agreement between two or more states regarding competing demands for a water resource that are beyond the legal authority of one state alone to solve. States administer water rights within their own political boundaries. This is a complicated process. The interstate compact is negotiated between states, adopted by their state legislatures, and then approved by Congress. After this is approved, each state may issue water rights to its share of water through normal administrative procedures. This usually occurs in Western states where allocations of water are traditionally done. An example is the Republican River Compact between Nebraska, Kansas, and Colorado (Corn and Soybean Digest, 2021).

Decreed rights are water rights determined by court degree. An example is the North Platte River Decree, between Colorado, Nebraska and Wyoming. The U.S. Supreme Court applied prior appropriation across state lines to determine water rights. Fully appropriate water sheds where water is coming into the watershed, is equal to water coming out of the watershed. These areas are closed to high capacity new wells (50 gallons a minute). Even though groundwater in Nebraska is owned by the public, you still need the landowner's permission to access that groundwater. Public ownership does not give the public an unlimited right of access. This is the case in most states. If you purchased adjoining property, you cannot horizontal drill a well and capture water from a neighbor's property (Corn and Soybean Digest, 2021).

Water access has been determined to be a basic human right by the United Nations General Assembly, and the Human Rights Council. The international human rights law exists to provide safe and clean drinking water to humanity. Thus, being a basic human right, promises have been made to do no harm to anyone's drinking water. The focus is on the basic needs of individual people (U.N., 2021). Yet, over a billion people in the developing world lack safe drinking water, and over three billion live without access to adequate sanitation systems. The result is water-related diseases such as cholera, diarrhea and typhoid. Furthermore, even in rich countries, some house-holds often to do not have safe water and sanitation facilities (Nehaluddin & Lilienthal, 2020).

The issues are, the core human rights instruments do not recognize a human right to water explicitly, but it can be implied in already recognized rights in documents that only have limited application. An independent right to water encapsulated in common law would have universal application. This would allow courts to directly apply the right to water, without having to refer to and be bound by other human rights, allowing for more protection. Three decades later, the independent right to water has yet to be clarified. The next step is to determine, if common law can be applied to international law. Moreover, to date, UN Member Nations are not in agreement over the right to water. This makes it more difficult but not impossible. The international community is starting to respond to the right for water more positively. Change is happening (Nehaluddin & Lilienthal, 2020).

We have focused on human use, agriculture, livestock, but what about the other users of water on our planet? According to National Geographic, an adult male elephant drinks up to fifty-five gallons of water in less than five minutes, and requires between eighteen to twenty-six gallons of water daily. When natural habitats are disrupted, elephant water supplies can diminish, along with food supplies and quality of life. On the Indonesian island of Sumatra, in 1986, fort-five hundred acres of natural forest were designated to the Balaj Raja Wildlife Reserve. This was specifically done to preserve elephant habitat. Today, only five hundred acres of forested land remains and the water sources the elephants and local people depend on are under threat. Over the last thirty years, so many acres of the reserve have been taken over by pulp and paper plantations where they have burned and deforested the natural habitat. Now electric fences protect their crops and block access to the natural corridors and water sources the elephants and locals have depended on for survival. Unfortunately, the elephants are forced to go closer to communities in search for water and food. This is a volatile situation (Global Green Grant Fund, 2018).

If we are to be a higher order in our thinking and decision making, then we have to be people who honor and respect all of humanity and all of the ones we share the earth with. We do this through education and holding others accountable, enforcing laws. Basic goodness and kindness can be taught, at least to the young.

REFERENCES

Aquaoso. (2021). *Current Water Rights*. Aquaoso. https://aquaoso.com/water-rights/

Bureau of Reclamation. (2021). *Water*. USBR. https://www.usbr.gov/main/water/

Case Text. (2021). *Bounds verse, the State of New Mexico*. Case Test. https://casetext.com/case/bounds-v-state-8

Corn and Soybean Digest. (2021). Deep Dive into Nebraska's water resources. Author.

Drought Monitor. (2021). *Drought Maps*. Drought Monitor. https://droughtmonitor.unl.edu/ConditionsOutlooks/CurrentConditions.aspx

EPA. (2021). *The Clean Water Act*. EPA. https://www.epa.gov/laws-regulations/summary-clean-water-act

Fletcher, A., & Fulford, S. (2017, Sep 2). Water my rights? Property law. *Financial Times*. http://0-search.proquest.com.libcat.uafs.edu/newspapers/water-my-rights/docview/1945406069/se-2?accountid=14481

Global Green Grants Fund. (2018). *Water for Elephants*. GGGF. https://www.greengrants.org/2018/03/22/water-for-elephants/

Immel, J. (2004). Best of the bar: Water Rights Law - Anne Castle, Castle enjoys navigating turbulent water rights law. *The Denver Business Journal, 55*(47). http://0-search.proquest.com.libcat.uafs.edu/trade-journals/best-bar-water-rights-law-anne-castle-enjoys/docview/233104778/se-2?accountid=14481

Matlock, S. (2010). Court rulings mixed on state's water-rights law. *McClatchy - Tribune Business News*. http://0-search.proquest.com.libcat.uafs.edu/wire-feeds/court-rulings-mixed-on-states-water-rights-law/docview/795518167/se-2?accountid=14481

National Center for Agriculture. (2021). *Water Law*. National Center for Agriculture. https://nationalaglawcenter.org/overview/water-law/

Nehaluddin, A., & Lilienthal, G. (2020). Right to Water as a Human Right: A Critical Overview of International Instruments. *Environmental Policy and Law, 50*(4-5), 299–308. doi:10.3233/EPL-200232

United Nations. (2021). *Human Rights to Water and Sanitation.* United Nations. https://www.unwater.org/water-facts/human-rights/

Water Education Foundation. (2021). *Riparian Rights.* Water Education Foundation. https://www.watereducation.org/aquapedia/riparian-rights

KEY TERMS AND DEFINITIONS

Abandonment: When a water right holder intentionally fails or neglects to use water for a period of time, despite the availability of water.

Adjudicated Water Right: A water right that has been perfected through a judicial adjudication clarifying and permanently establishing through decree the location, amount, use, timing, and priority date for the water.

Adjudication: A judicial proceeding by which vested or existing water rights, federal reserved water rights, and water rights awarded by the state under a permit system are integrated, and priority dates for all water rights are determined for an entire watershed or ground water basin. The process is referred to as a general stream adjudication.

Annual Peak Flow: The maximum instantaneous discharge from a stream. It is the highest annual discharge and includes both groundwater contributions and direct runoff.

Appropriated Water Right: A water right acquired through the prior appropriation doctrine, as provided by state law, as either a vestee, existing, or permitted water right. Often referred to as a "state-law water right."

Aquifer: A geological formation containing or conducting groundwater, especially one that yields a usable quantity of water.

Baseflow: The portion of the stream flow unrelated to direct storm runoff. It is the low flow of the stream that occurs through groundwater contribution to the channel.

Beneficial Use: The use of water for a purpose identified by the state as a beneficial use. Each state defines in law what uses are beneficial. They typically include consumptive uses such as agriculture, domestic, industrial, mining, recreation, etc. They may also include non-consumptive uses such as fisheries, riparian, aesthetic, recreation, etc.

Express Federal Water Right: When the federal government reserves public land and explicitly expresses an intent with regard to water.

Federal Reservation: A dedication of federal land to a specified purpose (for example, wilderness, national park, wildlife refuge, national conservation area, national recreation area, and many others).

Federal Reserved Water Rights: A water right reserved by the federal government (either expressly or impliedly) which supports the purpose of a federal reservation of land. Those purposes are specified in the Congressional legislation or Presidential proclamation that created the reservation.

First-in-Time, First-in-Right: A principle of prior appropriation doctrine. Every water right is recognized with a priority date. The earliest date is the most senior right, with the later dates a junior right to those preceding them. The most senior right is entitled to water first, followed by the next most senior right, and so on. When water supply is limited, the most junior rights may not be fulfilled.

Forfeiture: The water right holder has lost the water right due to failure to beneficially use the water for a specific period of time set by the state.

Groundwater: Water that collects beneath the ground surface in the pore spaces of soil, gravel, sand, or silt, in the pore spaces within permeable rock, and in the fractures or cavities of rock.

Hybrid Doctrine: Contains elements of both the Riparian and Prior Appropriation Doctrines in states that began using the riparian system but later changed to using the prior appropriation system. Consequently, some water use dating to very early development may be managed as prior appropriation rights, even though both are within the same state. Some states may manage surface water under the prior appropriations doctrine, while ground water is managed under the riparian doctrine.

Hydrograph: A graph showing the rate of flow (discharge) over time past a specific point in a river, or other channel or conduit carrying flow.

Implied Federal Reserved Water Right: When the federal government reserves public land for a specific purpose, and does not explicitly state an intent with regard to water, it implicitly reserves sufficient water to satisfy the purpose of the reserved public land.

Imported Water: The unnatural contributions of water to a stream. It occurs due to diversion of water from one system into another. This water in addition to the natural baseflow is considered together as the total baseflow.

Instream Use or In-Place Use: The use of water that does not require diversion. For example: retaining water in the stream to benefit fisheries, or the capture of water in a stock pond without diversion.

Junior Water Rights: Under the prior appropriation doctrine, water rights that fulfill after older rights (senior) have been fulfilled, but will not be fulfilled when the source is insufficient to support more senior rights.

Natural Flow Principle: A method of water allocation under the Riparian Doctrine whereby a landowner has the right to a natural water flow of undiminished quantity and unimpaired quality, and assumes a negligible withdrawal of water by any landowner.

Permitted Water Right: Water rights issued as a permit under state water statutes, which specifies location, amount, use, timing, and priority date for the water.

Phreatophyte: Plants with roots constantly in touch with inundated soils or ground water. These plants tend to utilized large quantities of water.

Prior Appropriation Doctrine: A method of allocating water rights whereby the first person to appropriate a quantity of water from a water source for a beneficial use has the right to continue to use the appropriate quantity of water for that beneficial use. Subsequent persons can appropriate the remaining water for their own beneficial purposes, provided they do not interfere with the rights of prior appropriators. Mostly applied in drier areas. Water flowing through a property in not owned as part of property ownership. This allows water to be moved to where it is needed most, and establishes a process to allocate water where there is not enough for all landowners. The water right then becomes appurtenant to the land on which the water is used.

Priority Date: The date used to determine senior and junior water rights. For a new permit applicant, it is the date the application is correctly filed. For vested water rights, the date of first beneficial use as determined by a court declaration. For federal reserved rights, the date Congress or the President established the federal reservation.

Reasonable Use Principle: A method of water allocation under the Riparian Doctrine whereby a landowner has the right to use water in a way that does not deprive or hinder other riparian users from other legal uses of water.

Riparian Doctrine: A method of allocating water rights whereby all landowners whose properties adjoin a body of water have the right to make use of the water. Mostly applied in wetter areas. If there is not enough water to satisfy all property owners, there is an equitable sharing of burdens in proportion to the property frontage to the water body. Water rights are appurtenant to the land and cannot be sold or transferred other than with the adjoining land. Water usually cannot be transferred out of the watershed if it diminishes other riparian landowners' rights.

Senior Water Rights: Under the prior appropriation doctrine, water rights that are fulfilled before more recent (junior) rights are fulfilled.

Storage Rights: The right to use stored water at any time after it has been appropriated and put into storage. The owner has complete control of the water upon its release from storage. Stored water may be released back into the stream for downstream use, and upon release is not subject to allocation based on priority date.

Surface Water: Water that collects on the surface of the ground in a stream, river, lake or wetland.

Vested Water Right: Often this is a water right established by the use of water before the state's water statues were enacted. These rights are confirmed through a court process called an adjudication. The court issues a decree recognizing the vested water right, identifying the date the use began, confirming the amount claimed is needed for the beneficial uses, and has been diverted continuously since the use began.

Water Right: A system of rights to allocate the use of water that travels or collects in streams, rivers, lakes, ponds, or underground, including the allocation of the water to storage. Water rights are property rights, but water right holders do not own the water itself, they possess the right to use it. Depending on the type of water law doctrine they may be attached to ownership of the land, or they may exist as a separate property right.

Well Construction Permit: A permit, required by most states or counties, to drill a well. The permit is used to protect aquifers from poor well construction techniques and may also prevent well-to-well interference, but the permit does not convey a water right.

Chapter 3
The Dowser

ABSTRACT

In the author's career, she has met many interesting people and acquired knowledge that she was very grateful for. She really does not know if dowsing is truly science or art. She does know that further investigation is needed to determine the science behind this. Her intuition tells her that there must be something unknown to us at this time, for dowsing to last a thousands of years, and have many scientists and engineers hiding their divining rods. We must remember that 100 years ago we thought things were myths when they were real. The unexplained is often science untested and unrevealed. Through autoethnography, this chapter explores the dowser.

"We must accept dowsing as fact. It is useless to work experiments to prove its existence. It exists. What is needed is its development." – Charles Richet, a Nobel Laureate

MEMOIR

I often did business with two brothers who owned a small water business where they bottled and sold private label water for events. They had their own printer to make labels and would customize labels for weddings, family reunions, birthday celebrations, and much more. This was an internet business

DOI: 10.4018/978-1-7998-9691-3.ch003

and very successful. It became so successful that in the late 1990's they needed to find more water and install another water tank. I had a stainless-steel water tank that we had outgrown and sold it to them very reasonably to help them. They invited me up to watch a well-known dowser find a new water source for them.

With my background, I was not convinced but went to watch because my friend has asked me. The gentleman was an older man, soft-spoken humble, and dedicated to his craft. He told me he really did not understand the science behind what he did but knew it was real. He had experienced over many years. He said when he was new at dowsing, he had some errors, but over the years had refined his craft, and for many years found he was accurate. He had a copper forked tool to do his work. He flags several places he felt would make great locations for new water well.

I asked how my friend and the dowser met, it all begins when a member of my friend's family became ill with cancer. The family doctor suggested that the power line that ran next to their house could be a source of cancer or cause cancer to grow. They researched how to move the power line, which was not an option because of the industry it fed, and the residential housing that depended on the power. The power line was new. In the process of researching the power line, they came across testimonials of cancer patients who used a dowser to re-direct the energy coming off the power lines away from their homes by hiring a dowser to do this.

After an exhaustive search, they found the dowser. He was highly recommended by many people in the state, even the state geological survey. I did raise my eyebrows at this because I had never heard of this, and frankly, was shocked. They measure the electrical feed coming off the power lines before the dowser re-directed the power and after the dowser redirected the power. There was a significant difference. From there they hired the dowser to find water and did every time, and the water source was prolific.

The geoscientist in me said, of course, our state's groundwater is vast, and it is not hard to find a good source of groundwater. However, their bottling plant was on the side of a Mountain and sat on bedrock. Bedrock sources mean that water is usually deep underground and comes to the surface on its own pressure or is redirected underground. They are not as easily located as a sand and gravel groundwater source, where water is held is a saturated zone (like a giant sponge) and is prolific. The dowser was accurate every time.

Of course, I felt this was either chance or a form of science that was unknown to me. I visited with the dowser and started asking him questions. I learned this art was generational and had been passed down to him from his father, his father from his grandfather, and went back in his family several generations. I asked if there was any genetics that they shared, and he said there were several unusual traits they all shared. They all had an allergic reaction to gold, so none word a gold wedding ring or watch. They could not even carry a gold pocket watch. He said he felt it was a neurological response because everyone male in his family that dowsed felt a tingling or slight shock go through their body when dowsing. This often goes down their back. When dowsing electrical lines, he often he could see a blue light coming out of the ground and others nearby could see it too, and the shocks to the body were much stronger than when looking for water. This seems to match the previous research discussed in this chapter.

The dowser and I developed a great friendship. He was a delightful person, a gentle engaging noble man. He passed away a few years after I met him, I went to his funeral, and the church was full of adoring people.

The next encounter I had with an experienced dowser was a few years later, in another state, and working with a family who was prominent in the area, well established, and a pleasure to work with. They hired a dowser who had worked with their family for many years locating new wells for the growing apple orchards.

This man was very nice but completely different from the first dowser I worked with. He was a tall imposing man, a gregarious personality, but became very serious and focus when working. He flagged areas of potential water. According to the landowners had always been successful. He begged me to spot water for me, but I just could not go there, especially at that time with company money. I probably would have been laughed at and ridiculed. I had my consultants, and my normal plan of action depended on a thorough investigation into land uses going back sometimes to the 1600s. We would look at the U.S. Geological maps, and scans, and determine where to drill. We were never wrong.

At break time I asked him questions, the same questions I asked the previous dowser. He gave me almost the same answers. He was allergic to gold. Dowsing was generational in his family going back through his ancestry as far as he could remember. His words, "it was in his blood". He only dowsed for water and had no desire to dowse for minerals, hidden artifacts or move electrical currents, but he did know dowers that did all these things. He was firm when he said this is a craft, one must develop this craft to be good at it.

In my career, I met many interesting people, and learn knowledge that I was very grateful to acquire. I really do not know if this is true science or art. I do know that further investigation is needed to determine the science behind this ancient art of dowsing. My intuition tells me that there must be something unknown to us at this time behind this, for it to last thousands of years, and to have so many scientists and engineers hiding their divining rods. We must remember that one hundred years ago we thought things were myths when they were real. The explanation is often science untested and unrevealed.

DOWSING

The science of water dowsing has been around for hundreds if not thousands of years. It is practiced by using a forked stick, rod, pendulum, or similar device to locate underground water, minerals, or other hidden or lost materials. Many people have referred to dowsers as diviners or water witches. Most of us the traditional forked stick, which may come from a variety of trees, including the willow, peach, and witch-hazel. In the more traditional method, the dowser holds each side of the forked twig with the palms upward. The bottom or butt end of the "Y" points skyward at an angle of about fort five degrees. The dowser walks back and forth over the area to be tested. When they pass over a source of water, the butt end of the stick is supposed to rotate or be attracted downward (USGS, 2020). Due to the expense of test drilling many homeowners and small businesses prefer to try a dowser first.

The Ancient art of dowsing has been practiced throughout millennia, although the names used to identify it may have changed in different cultures and eras, the techniques have not. In 1949, a party of French explorers (while searching for evidence of lost civilizations in the Atlas Mts. of North Africa) stumbled upon a massive system of caverns known as the Tassili Caves. Here the walls of the caves were covered with marvelous pre-historic paintings. Among the many fascinating wall murals, not only did they locate an art gallery devoted exclusively to the depictions of spacecraft and ETs, but they also found a remarkable huge wall painting of a dowser, holding a forked branch in his hand searching for water, surrounded by a group of admiring tribesmen. These wall murals were carbon dated and found to be a least 8000 years old.

In China, there is an etching of Chinese Emperor Yu who ruled China 2500 years ago, and in his hands, he holds a rather bulky turn-pronged device that resembles a dowsing device. Furthermore, many passages in the Bible allude to dowsing. Moses and his son, Aaron, used a dowsing device referred to as "the Rod" to locate and bring forth water.

In the 1400s, dowsing as we think of it today was called "Virgula Devine" in Latin which meant dowsing with the rod shape. In Germany, during this period, dowsing devices were used extensively by miners seeking mineral ore. According to Christopher Bird, author of the classic book, "The Divining Hand", no one is absolutely certain of the origin of the verb to dowse". But it seemingly made its first official appearance in 1650 in an essay written by the famous English Philosopher John Locke whose noble writings inspired the framers of our own Declaration of Independence and The Constitution of the United States. In his essay, Locke wrote that using the dowsing rod, one could devise or discover water and precious minerals (such as gold & silver and mineral ore) Locke appropriated his phrase from the long-dead English west country language of Cornwall – where in Cornish Dewsys meant "Goddess", and "Rhod" meant tree branch, and from which he "coined" the phrase – Dowsing Rod. In the 1700s and 1800s in England, Germany, and France various books on mining and engineering were preferred.

London, England's 1912 edition of "Mining Magazine" published the first translation of a Latin Opus into English. It was called "On Metals" in praise of dowsing which had been first published 356 years earlier. The translators were a professional American mining engineer and his wife. The engineer later gained fame and won notoriety as the 31st President of the U.S. — Herbert Clark Hoover. As Chris Bird notes, "God knows, had President Hoover been an expert dowser himself, he might have predicted, and therefore, prevented the great stock market crash of 1929" (ASD, 2019).

When looking at dowsing and the science of water there are some explanations. One is that groundwater is very prevalent and close to the surface. You can run a household on a dug well, maybe not in a drought, but in normal rainfall years. In a region with adequate rainfall and favorable geology, you can find water easily. Many dowsers appear to be successful, but they are not able to understand the depth, quantity, and quality of the water they are located. It takes science, geologic, hydrologic, and geophysical knowledge to thoroughly understand the science. The area must be tested and studied to understand the water source and the watershed.

Ten out of Twelve water companies in the United Kingdom rely on dowsers to find water sources, water lines, and other underground materials. The companies found that dowsing is done by individuals and cost-effective, it may not be one hundred percent effective, but is most of the time (LaPage, 2017). The U.K. is not the only country where dowsing methods are being used to search for water. As NPR's Julie McCarthy found last year when investigating a drought In India, the local farmers were so desperate they hired a dowser to find water. After several years of drought, the Bendsura Project a large dam and reservoir are empty. This supplied twenty-eight thousand villages. Desperate for water farmers and homeowners have drilled wells. They hired a dowser to help them place their wells. The dowser uses a copy diving rod. As he finds water, he moves to other directions to see if there is a channel. The dowser often will take up to three hours traversing a plot of land to ensure he has found the richest vein of water. The dowser worked for four hundred families with an eight percent accuracy rate (McCarthy, 2016).

Society of Dowsing

The American Society of Dowsers was formed by 11 men in 1961. The American Society of Dowsers, they have the largest organization of dowsers in the world. Many are in the United States, they are the most skilled dowsers, ranging from water dowsers to seekers of minerals and lost objects, energetic dowsers, esoteric dowsers, and a great deal more (ASD, 2019). Their mission is: (ASD, 2019).

To support, encourage and promote dowsing and dowsers in a manner consistent with the highest standards of personal integrity and behavior; to provide dowsing education and training to dowsers and non-dowsers alike to bring them to a level of proficiency they are comfortable with; to promote and foster communication and fellowship among all persons in any way interested in dowsing."

- **Dowsing Theories** – disseminate knowledge of dowsing (water witching, the discovery of lost articles or persons, and related para-psychological phenomena), development of its skills, and recognition for its achievements.
- **Quarterly Journal** – The society issues the quarterly journal the American Dowser, which presents a representative cross-section of this highly diverse material.
- **Research Led** – American Society of Dowsers is led by the research of dowsing theories and techniques.

"In hundreds of cases the dowsers were able to predict the depth of the water source and the yield of the well to within 10 percent or 20 percent," says Hans-Dieter Betz, a physicist at the University of Munich, who headed

Figure 1. Diving rod (1999).

the research group. Even at this rate, they were not able to change minds. The German researchers did not give up and found subtle electromagnetic gradients within natural fissures and water flows create changes in the electrical properties of rock and soil. The German researchers theorized that these low-energy gradients respond by wagging the forked sticks and bent wires. There is ample evidence that humans can detect small amounts of energy. Biologists have found nonvisual electric and magnetic sensing organs in creatures from bacteria to birds. The German researchers have no theories on how dowsers come by their skill, but two things are certain after 10 years of research, is that a combination of dowsing and modern techniques can be both more successful and less expensive than relying on only modern techniques (Popular Mechanics, 2004).

SCIENCE OF DOWSING

The "Bio-Physical Method" (BPM) is now being considered as a scientific explanation for dowsing. This idea has a small following in the mining industry and other sectors of commerce. This is being hotly debated, many tested cases and success is found to be often enough not be dismissed as chance or coincidence. Some of the explanations focus on the assumption that dowsers are more sensitive to the Earth's magnetic field and pick up on small changes in it caused by water or metal underground. The theory suggests that when a dowser detects a magnetic anomaly, it sends a signal to the dowser's muscles, which experience a small contraction, causing the rod to deflect. Other researchers suggest that dowsers respond to the anomalous electrical fields in the vicinity, which they say affect their blood pressure and their pulse rates (McWilliams, 2020).

Researchers at the California Institute of Technology (Cal Tech) (2019) recently identified tiny crystals of magnetite in human brains. This iron oxide which helps bird navigate, maybe how humans have an unconscious and rudimentary directional sense. Dowsers believe this discovery could be the basis of their gift. Some individuals may have a higher level of magnetite than others. Critics remain adamant that successful dowsing is more based on the geology of the area than the ability to detect magnetic or electrical anomalies in the vicinity (McWilliams, 2020).

Researchers at Cal Tech found that many humans unconsciously can detect changes in the Earth's magnetic fields. The study was led by geoscientist Joseph Kirschvink, neuroscientist Shin Shimojo and nanoengineer Ayu Matani determines there was experimental evidence that human brain waves respond to controlled changes in Earth-strength magnetic fields, which is the first evidence of a new human sense: magnetoreception. Kirschvink and Shimojo built an isolated radiofrequency-shielded chamber and had participants sit in silence and utter darkness for an hour. While the participants were in the chamber, they shifted the magnetic field silently around the chamber and measured participants' brain waves with electrodes positioned at sixty-four locations on their heads (Kirschvink, Shimojo, & Woodard, 1992).

Thirty-Four humans from a wide age range and a variety of ethnicities participated in the experiments. Many of the participants experienced changes in their brain waves, that correlated with the changes in the magnetic field around them. As researchers tracked the alpha rhythm in the brain, which occurs at between 8 and 13 Hertz and is a measure of whether the brain is being engaged or is in a resting mode. When the brain is unengaged, the alpha power is high. Moreover, when the brain is unconscious or conscious and gives attention to something the alpha power drops. Several other sensory stimuli like vision, hearing, and touch are known to cause abrupt drops in the amplitude of alpha waves in the first few seconds after the stimulus (Kirschvink, Shimojo, & Woodard, 1992).

Basically, the experiments determined that in some participants the brain appears to actively procession magnetic information and reject signals that are not natural. When the magnetic field pointed upward during the experiments, there were no corresponding changes in the brain waves, this is because the magnetic field normally points down in the Northern Hemisphere. It is seeming like the brain is ignoring signals that are obviously wrong (Kirschvink, Shimojo, & Woodard, 1992).

One of the challenges in early attempts was the difficulty of making sure that those changes in brain waves were, indeed, correlated to the magnetic field and not to some other confounding effect, such as the coils generating around the chamber created an audible hum this might interfere with the alpha power in participants. To remedy this, the researchers the chamber was not only pitch black and isolated, the copper wires for altering the magnetic field were wrapped and cemented in place in duplicate with each coil a pair of wires, not single. As the current is directed through the pairs of wires, the magnetic field in the chamber is altered. This entire experiment was controlled by computers. Results were processed automatically with no subjective input. (Kirschvink, Shimojo, & Woodard, 1992).

Aristotle described the five basic senses as including vision, hearing, taste, smell, and touch," says Kirschvink, co-corresponding author of the eNeuro study by Marilyn Van Wingen Professor of Geobiology. "However, he did not consider gravity, temperature, pain, balance, and several other internal stimuli that we now know are part of the human nervous system (Kirschvink & VanWingen, 2019).

By developing and demonstrating a methodology for human testing, researchers determined that this can be the roadmap for further research into the presence of highly evolved geomagnetic navigation systems in man. Surprisingly, the question is, can this research uncover the ability of the dowser to find water and materials unground?

Dowsing is often described as the "Art of Knowing" because dowsers search for the unknown. The research is cut and dried, often confusing to mainstream science, but cannot be ignored due to the reality of the surprisingly reproducible phenomenon. It is very difficult to determine which perimeters are and which results are relevant. Electromagnetism is in a stage of early investigation, and we are just beginning to understand science. We may have to start considering evidence base dowsing, where the dower walks over a site, the rods respond, and the dowser moves on to remote dowsing. Research has determined that around ninety percent of the population can do this, but it is ignored by the scientific community. It could be based on basic human neurophysiological responses to an unknown stimulus (Miller, 2021). We must keep in mind that dowsing is widely used as a start, which leads to more scientific techniques for gaining information about a site.

Geologists and engineers are very pragmatic, and not given to techniques that usually do not work. So, it is surprising that dowsing has been used for so long and so often. In the mid-1600s a scientific report for dowsing was presented to the Royal Society. Recently, the Cornish Tim industry hired German dowsers to come to train their people to dowse for tin. Many engineers have divining rods as part of their toolbox (Miller, 2021). Mainstream industries from drug, auto, and food manufacturers have used dowsers to find new water supplies. True water scientists would say they do not, or never admit if they did.

In 1971 Chadwick and Jensen of Utah University investigated if novice dowsers could replicate each other's results. Out of one hundred and fifty participants, most were students and staff of the university, and to the surprise of Chadwick and Jensen, they were able to replicate each other's results. Chadwick and Jensen are electrical engineers and compared their results to a recent magnetic survey. They determine that there were some correlations between the results, due to more of the dowsing occurring where the magnetic field gradient was over 1.6mTm than when the gradient was less. Moreover, these results are not as promising as the result that the novice dowsers were able to achieve the same results in the same places (Hensen, 1982). Similar results were obtained by a research group working for the British Society of Dowsers in 1997.

The most interesting research was performed by the German Government and the University of Munich. This was a ten-year study in the 1980s on the application of dowsing in arid regions. They monitored experienced dowsers for the study. The results were surprising. In six hundred and ninety-one test drillings in Sri Lanka, dowsers achieved an overall success rate of ninety-six percent. With conventional science, the success rate was only thirty to fifty percent successful. These results supported other research being performed by Soviet geologists in the 1960's and 1970's, where dowsers achieved a ninety-one point five to ninety-four percent success rate in siting water wells (Miller, 2021).

Experienced dowsers can detect almost anything at any distance, disturbed ground, oil, minerals, as well as water. However, wet sands and clays seem to reduce the penetration of the dowsing ability through the ground. The experienced dowser learns to focus and does not get confused by parallels. Parallels, such as a pipeline, can confuse a novice dowser. The spacing of these parallels oscillates suggesting a possible link with the normal daily variation of the earth's magnetic field (Miller, 2021).

Dowsers that are interested in geopathic stress have identified zones that are beneficial and harmful to human health. Most of this work has taken place in Europe where cancer deaths were above the national average. A group of doctors and dowsers mapped zones of severe geopathic disturbance in five hundred and sixty-five houses and correlated the zones with death records. The study concluded that earth's radiation typically detected by dowsers was the prime cause of the most serious diseases (Hansen, 1982).

Many researchers have universally accepted that dowsing is a neurophysiological response and that diving rods are a mechanical amplifier of small measurements to the body. Rural dowsers seem to use the traditional forks and sticks, whereas more professionals use metals. Medical professionals seem to prefer pendulums for a greater range of potential responses. A range of neurophysiological responses has been documented. A tingling in the arms or solar plexus. A shudder down the back or clicking in the ears. Novice dowsers have even reported vomiting. These reactions could be a response to stimulation of the nervous system (Miller, 2021).

Just because the rod has moved, the result in not always easy to interpret. The dowsing technique has always been to pick up subtle responses, however, experienced dowsers are able to recognize and eliminate this effect. When gathering data, it is very important that the dowsers have no possible clues on the location. Some experiments have dowsers blindfolded. It could just be that the rods act as sensitive amplifiers if they are held virtually horizontally, but without visual feedback, it may be difficult to maintain this (Miller, 2021).

Interpreting the results may complicate experiments. A delay may occur between the dowsing and brain which would create a small reflex in action. The size of the dowsing response seems to be dependent on the rate of change of the signal. A dowser can produce a bigger response by walking faster over the zone of interest. In addition, the same effect is often seen in the dowsing response with the rate of change of the magnetic field surrounding the dowser. Dowsers find their ability to pick up and interpret these subtle signals to improve with experience (Miller, 2021).

Dowsers use a variety of techniques to determine the depth of water or to focus attention. The mental state and stability of the dowser are important. For the best results, a dowser should be calm, relaxed but able to focus, almost in a meditative state. Geoff Brooks, a psychologist, and dowser has reported measurements showing an increase in theta wave while dowsing. Dowsing may relate to the electrical characteristic of the body. A Dutch geophysicist (Tromp, 1940) performed a two-year study on dowsers and found that they have a lower palm-to-palm skin resistance than nondowsers. Tromp claimed that washing hands reduces skin resistance and may temporarily increase dowsing sensitivity. Furthermore, there was a significant change in the skin potential measured between the wrist or dowsers, when passing over a dowsing zone (Tromp, 1949).

In 1978 a demonstration of the induction of a dowsing response by an electromagnetic field was carried out. This was a double-blind trial on the response of fourteen experienced dowsers to a low-power high-frequency electromagnetic field, switching on and off randomly. The results were significant. Out of six hundred and ninety-four trails, the dowsers scored ninety-five percent hits in comparison with the chance level of fifty percent. Harvalik concluded that the dowsing sensors are in the region of the kidneys and the pineal gland in the brain (Harvalik, 1978).

It is intriguing to think that the neural network within the body is affected by a wide range of electromagnetic influences including magnetic and electric fields as well as electromagnet radiation. Thus, the dowser's "Art of Knowing" involves learning to recognize and respond (Miller, 2021).

Dowsing could be a bridge based on grounded complex interactions between neurophysiology, and electromagnetic zones. As research becomes more advanced, and electromagnetism is more explained. Then dowsing may be proven to have a science core.

REFERENCES

Allen, K. (2020). *Is there Any Scientific Evidence for Dowsing?* Science Focus. https://www.sciencefocus.com/science/is-there-any-scientific-evidence-for-dowsing/

American Society of Dowsers. (2019). *History and Mission*. American Society of Dowsers. https://dowsers.org/

Graves. T. (1976). *Dowsing Techniques and Applications.* Turnstone Books.

Hansen, G. P. (1982). *Dowsing*: A review of experimental research. *Journal of the Society for Psychical Research, 51*(792), 343–367.

Kirschvink, J., & Shimojo, S. (2019). *Evidence of Human Geomatic Sense.* Cal Tech. https://www.caltech.edu/about/news/evidence-human-geomagnetic-sense

Kirschvink, J., & VanWingen, M. (2019). Scientist Develop Experiments that Human Brain Waves Respond to Changes in the Earth's Magnetism. *Science Daily*. https://www.sciencedaily.com/releases/2019/03/190321083637.htm

Kirschvink, J. L., Kirschvink, A. K., & Woodford, B. J. (1992, August). Magnetite Biomineralization in the Human Brain,. *Proceedings of the National Academy of Sciences of the United States of America*, *89*, 7683–7687. doi:10.1073/pnas.89.16.7683 PMID:1502184

LaPage, S. (2017). *UK Water Companies Use Dowsers*. NPR. https://www.npr.org/sections/thetwo-way/2017/11/21/565746002/u-k-water-companies-sometimes-use-dowsing-rods-to-find-pipe

Maby & Franklin. (1939). *The Physics of the Divining rod.* Bell.

McWilliams, B. (2020). Scientific Theories to Explain the Dowsing Art. *Irish Times*. https://www.irishtimes.com/news/scientific-theories-to-explain-dowser-s-art-1.245860

Miller, A. (2021). *A Review of Dowsing*. Scientific & Medical Network. https://explore.scimednet.org/index.php/dowsing-a-review/

Popular Mechanics. (2004). Dowsing Data Defy the Skeptics. *Popular Mechanics.* https://www.popularmechanics.com/science/a3199/1281661/

Tromp, S.W. (1940). *Psychical Physics.* Elsevier.

Chapter 4
The Zinc Mines

ABSTRACT

This is an example of how uninformed the public is, especially before the Clean Water Act. Pregnant mothers drinking contaminated water gives the same contaminate to their babies or unborn children. Probably what saved them from dying early was that the spring that was piped into their house was the headwater to streams in the area. When the authors tested the water at the spring, it still had moderate levels of sulfides, thus the smell she was smelling at the spring. In addition, she found levels of cadmium and arsenic. The dust from the open pits is toxic and may even be radioactive. Through autoethnography, this chapter explores the effects of zinc mines on water supplies.

The earth, air, land, and water are not an inheritance from our forefathers but on loan from our children. So, we have to hand over to them at least as it was handed to us. – Gandhi

MEMOIR

One time I received a phone call and a very nice man on the phone relayed to me that he had a beautiful spring on his property. Excited, my young intern and I set off to the mountains of Pennsylvania. As always, I researched

DOI: 10.4018/978-1-7998-9691-3.ch004

the U.S. Geologic maps. To myself, I thought this could be promising depending on how much of the recharge area the family owned. We packed our equipment and coolers full of bottles to send off to various labs. Early on in our relationship, and because so many of the calls we were getting we not feasible, my intern and I developed a code. If I asked for the small cooler, it meant I am not interested in the site but would send the samples through the mass spectrometer at work to give our hosts something for their own knowledge and to help them manage their property.

Our destination was north of Allentown Pennsylvania and our manufacturing plant. We were driving through a beautiful area; it was a nice sunny day and a great day to be doing fieldwork. Then, as we came to the top of a small mountain, we were blown away. I had a hard time not stopping on the road to stare. I pulled over and rolled down my window and a foul stench hit my nose. We had entered the land of zinc mining. I knew the devastation of zinc mining was bad, but seeing it firsthand is entirely different. Tears came to my eyes, and not because of the stench. The mountain was stripped bare, no trees or vegetation existed. I turned to my intern, and said, "This looks like the mining in the Lord of the Rings". These were open pits. Open pit mining degrades the land, leaving it barren, dust is full of poisonous gasses, water in the area is polluted, and creates a complete loss of the ecosystem. There was no activity, but I started to wonder what was ahead of us.

We continued, and came out of the mine area, and came into the vegetated and forest area again. We reached our destination and entered a long driveway to a cute cape, white with black shudders. It was a picture-perfect setting. Except for one thing.... Loads of dog poop on the ground. Not just a little, it covered the ground. I hate dog poop; it really makes me sick. On little note here. Once, I was with a friend walking her dog, in Rye, New York where the laws are very strict. Your dog poops and you must pick it up then. She went into the local small store and left me with the golden lab, and he decides to do his business right there in front of the store. She left me the baggies to pick up the poop, but as I was trying to do so, I began to dry heave. It was not pretty! I guess it caused quite a commotion in the store, everyone was looking out the windows to witness this sick woman trying to pick up dog poop with a baggie. I appreciated the fact some found humor in it. The dog and I did not! So, you can get the picture, of how bad dog poop is for me. At this time, I thought dog poop and zinc mining dust was the worst smell I ever encountered. I was so wrong. That is in another chapter.

Now back to my story. I stepped out of the car, and so did my intern and we both stepped in a big pile of dog poop. At the same time, a man came out of his house with four dogs. He had the sweetest smile and was very welcoming, the dogs were too. I smiled the best I could do and looked around for a place to wipe the dog poop off my boots. (Thank God I had boots on) There was dog poop everywhere, I mean everywhere. Needless to say, I began to vomit, which in turn started my intern to vomit. The man stopped in his tracks with wide eyes. I proceeded to tell him, that I have a strong aversion to dog poop. You could tell he felt bad for us but was really at a loss as to what he could do. I got paper towels out of the trunk and found a grassy place to wipe my boots.

It was then, after taking some water, that I could focus on my host. There was a movie that came out in 1991 called "Nothing but Trouble", it starred Dan Aykroyd as judge Alvin Valkenheiser, well this man looked just like the judge Alvin Valkenheiser. This was 1994, so it was not that long ago that the movie came out. I looked over at my intern, and she was wide-eyed and quiet. Despite the initial impression, he was a very sweet and kind person. The kind of person you just want to help and make happy.

We proceeded to his spring which was behind his house, up the top of a mountain. By all appearances it was beautiful. This amazing prolific cold spring water bubbling up from the ground. When I got close enough, I could pick up a smell that I did not like. I turned to my intern and asked her to get the small cooler. We would take samples and put them through the mass spectrometer at our own lab. While she was gone, I talked to the man in dept about how much property he owned and what bordered his property. He owned a large parcel of 500 acres. The federal government-owned one side of his border, and a corporation owned the other. He was so proud that he never sold off his land for development. He had lived on the property with his parents his entire life until they both died of cancer in their seventies. He now lived with his beloved dogs.

I told him we would test his water, and I would stop at the town and county offices to check out the land uses around him. Then we would let him know the results. I usually spent an enormous amount of time in town or county offices going back through deeds and land uses as far as records would take me. Often going back to the 1600s and finding deeds with the marks of native people. In Pennsylvania, I once found property once owned by William Penn. Even when the water quality comes back pristine, I would perform exhaustive research on the property.

We said our goodbyes and headed down the road to pull over into a grassy place to clean off our boots once again. I tried to get my brake and gas pedal clean with paper towels, all the while a cold sweat overcame me. We headed into the nearest town to wash the car and go to our hotel rooms. I had to take a shower, I could smell dog poop, and it was still making me sick. My intern felt the same. After getting refreshed we went to the town office to investigate. It did not take long to learn that a superfund site was on the other side of the mountain. We started to learn the extent of zinc mining in the area. My heart broke for this man and his family, and I thought "how many other lives have been impacted".

We went back to the plant, and I called the Delaware River Basin Commission. The Delaware River Basin Commission was established in 1961 by the federal government to create an independent legal authority to oversee and manage the Delaware River Basin. The commissioners overseeing the commission are the governors from four states Delaware, New Jersey, New York, and Pennsylvania. The staff consists of highly qualified scientists and administrators. In addition, the commission has advisory committees from the communities they serve and top scientists. There are advisory committees for both water quality and toxic waste (DRBC, 2021). They confirmed my fears. I asked if there were any grants to help our friend install water purification in his home. They said they would reach out to him. In the meantime, I sent a month's worth of water to him for drinking purposes.

ZINC

Zinc is an element commonly found in the Earth's crust. It is released to the environment from both natural and man-made sources. However, releases from man-made sources are far greater than those from natural sources. The primary man-made sources of zinc in the environment (air, water, soil) are related to mining and smelting involving zinc and the use of commercial products containing zinc. Releases to soil are probably the greatest source of zinc in the environment and by-products of open pit mining and smelting being released into the air. The most important sources of zinc in the soil come from discharges of smelter slags and wastes, mine tailings, coal and bottom fly ash, and the use of commercial products such as fertilizers and wood preservatives that contain zinc. Although zinc usually remains adsorbed in the soil, leaching has been reported and causes contamination. Zinc does not volatilize from soil or water but is deposited primarily in sediments through adsorption, precipitation, and airfall (CDC, 2019).

Figure 1. Zinc
(Green speck, UK, 2011).

Zinc has been found in concentrations at hazardous waste sites and industrial areas. What is concerning is that zinc has been found in air, surface water, groundwater, and soil. Zinc has been found in human food and is often considered a good supplement to prevent colds and viruses. The danger is taking too much zinc and overdosing on it when using it as a supplement. Certain populations are and have been exposed to high concentrations of zinc, and those groups are usually found in mining or industrial areas. People who work in coal mines, smelting operations may be exposed to excessive amounts of zinc poisoning (CDC, 2019). If you love to eat mussels and oysters you may have been exposed to elevated levels of zinc.

Figure 2. Zinc mine
(Delaware River Basin, 1994).

Manufacturing and smelting facilities are required to file a yearly report to the EPA (EPA, 2020). This report is called the *Toxic Release Inventory* and is only a requirement if the company employs ten or more full-time employees if their facility is classified under the Standard Industrial Classification codes 20-39 and if their facility produces, imports or processes more than 25,000 pounds of a Tri chemical or otherwise uses more than 10,000 pounds of a Tri chemical in a calendar year. The Tri chemicals are a list of chemicals that cause one or more of the following: (CDC, 2019).

- Cancer or chronic human health effects
- Significant adverse acute human health effects
- Significant adverse environmental effects

Zinc is one of the most widely used metals in the world. Major industrial sources of zinc include electroplating, smelting and ore processing, and drainage from both active and inactive mining operations. Furthermore, zinc is an important component of brass, bronze, die casting metal, other alloys, rubber, and paints (CDC, 2019).

Volcanoes release zinc into the atmosphere. The estimate is around 35, 899 tons yearly. Mount Pinatubo released an estimated 800,000 tons of zinc into the atmosphere in 1991 (Garrett, 2000). Other biogenic emissions and sea salt sprays are estimated to be four hundred forty metric tons a year. Humans release zinc into the atmosphere by several means. However, since the 1970's improvements have been made to facilities and zinc emissions have decreased around seventy five percent for air and eighty five percent for water (CDC, 2019).

Erosion is the largest contributor of zinc in water. Soil erosion resulting from human activities accounts for over seventy percent of soil loss. Geologic or natural erosion is around thirty percent of soil loss (USGS, 2020). Urban runoff, mine drainage, and industrial and municipal effluents are smaller but more concentrated sources of zinc in water (CDC, 2019). Accidental zinc discharges to water are associated with smelting and refining operations. Drainage from active mining areas is considerably less than from inactive areas because of the disposal methods currently employed. Hazardous waste sites, in which zinc has been improperly disposed of, are additional sources of the element. Metals, such as zinc, also enter estuaries from many natural and manufactured sources. Zinc is often present in soils and grasses because of atmospheric aerosols. Furthermore, approximately 22,000 tons of zinc are used in fertilizers each year in the United States (EPA, 2000). In addition, zinc is an essential nutrient that is present in all organisms.

MINING IN PENNSYLVANIA

Currently, Pennsylvania has one thousand two hundred and five mines listed by the United States Geological Survey, mining iron, zinc, lead, manganese, and copper. Coal is mined in the Northeastern part of the state. Fifty-nine mines have an open pit, outcrop, or isolated drill hole – known as an occurrence mine. Pennsylvania has sixty-three prospect mines. Currently, according to the U.S. Geologic Survey, there are 1,020 mines in production (USGS, 2020).

Zinc Mining in Pennsylvania

The birth of Pennsylvania zinc mining started in 1845 in the Saucon Valley near the village of Friedensville in Lehigh County. A man by the name of Theodore William Roepper the first professor of geology and mineralogy at Lehigh University, determined that the material to be mined was hemimorphite, known as a calamite, a zinc silicate, and smithsonite or zinc carbonate. If smelted by combining it with native copper, it produced brass. In 1847 shallow exploration shafts produce zinc ore. Samuel Wetherill who operated the Philadelphia chemical works was searching to replace lead in their paint, and zinc could be an alternative, He filed a patent. At the same time, another man patented the process for collecting zinc ore particles from furnace smoke in long muslin bags.

After mergers and new processes, the heyday of the Friedensville mines was from 1853-1893. Mines were continually added. However, groundwater started pouring into the mines and proved a great risk to the miners. They did not have the equipment to sufficiently pump the groundwater out of the mine. In the 1930s and 1940s new technology allowed groundwater to be pumped out of the mine. This pumping capacity actually caused a severe drought in the 1950s. Many of the springs and wells dried up in the area. In the 1970s pumping capacity increased again. By the 1980s zinc prices were falling, and the mines shut down.

In 1984 water sampling in the area contained small amounts of cadmium and arsenic. South of the mine, several housing subdivisions had been developed and light industrial parks had risen. In addition, a large shopping center had been constructed. The Palmerton Zinc Pile Superfund site is in the borough of Palmerton, Pennsylvania. It is the site of two large former zinc smelting operations. In 1983, the Environmental Protection Agency (EPA) placed the site on the Superfund program's priorities list (EPA, 2020).

The EPA determined that the smoke stacks from the New Jersey Zinc Co. created airfall deposits throughout the 20th century in the region surrounding Palmerton, PA. This led to the destruction of the forest ecosystem along the neighboring Kittatinny Ridge and metal contamination in the town and surrounding area. Contamination is still leaking into the groundwater and the environment even after this operation was shut down in 1983. Furthermore, it was determined the Superfund site was emitting airfalls, and testing showed levels were hazardous for young children and pregnant women. This area is uncomfortably close to the spring we went to see. I think about that sweet man, drinking and breathing contaminants his entire life.

In February of 2021 federal officials settled a lawsuit with American Zinc Recycling over their Palmerton plants. According to the Morning Call (2021), the following settlement was reached:

- American Zinc Recycling Corp. will make $4.3 million worth of improvements to its Palmerton plant in a settlement ending a lawsuit by state and federal authorities over violations of environmental laws, including emissions of toxic lead dust.
- The company, formerly known as Horsehead Corp., will also pay a $3.3 million civil penalty, which will be divided between Pennsylvania and the U.S. government.

According to the EPA violations of the Clean Air Act and permits issued by the Pennsylvania Department of Environmental Protection in 2017. For the last one hundred and twenty years, the Palmerton plant has been used for zinc smelting and processing. Pollution sources of zinc, copper, lead, cadmium, and chromium pollution cause deforestation of the Blue Mountain north of the Leigh Gap. The site and surrounding area was designated a Superfund site in 1983. Since 1986, the site has been used only for processing zinc and other material which is now home to a facility that reclaims zinc and other materials from a byproduct of the steel industry.

According to EPA's acting assistant administrator, lead emissions from the plant have continued to be a problem for the area. The EPA determined that the settlement would address emissions of lead dust that are a public health risk to families living as close as three miles away. In 2019, a Palmerton day care center tested high for lead. The findings were the result of testing performed after a federal study found high lead levels in the air, causing health risks. Free blood testing by the state Health Department found that twelve out of two hundred and sixty-seven people, including three children, had elevated lead levels. The borough has not been required to perform remediation at the town park. The Pennsylvania Department of Environmental Protection said the park was safe. In addition, the lawsuit alleged that there were violations of the Clean Water Act, exceeding limits for cadmium and zinc emissions, wastewater discharge, and violations of stormwater management and spill prevention regulations (Hall, 2021). I am grateful to see fines, but when will it end?

MINING

Open pit or strip mining takes place on the surface. Forty percent of mining occurs at the surface of a mining site. The surface environment is destroyed and polluted during the mining process. One of the major benefits to open pit mining is the increase in efficiency over deep-shaft mining procedures. Mining that occurs at the surface has no narrow tunnels and shafts that affect the rate at which ore can be extracted. Open pit mining makes it easier for surveyors to analyze the potential ore yield and avoid safety hazards. It takes larger extraction vehicles, which increases the yield.

All this comes at a great cost because open pit mining virtually eliminates any biological life at the surface. Vegetation is stripped and the site is left barren. Without replanting and restoring the ecosystem, an open pit mine will take decades to recover. In addition, there are additional risks. The abandoned pits have steep vertical walls, and the structure is not stable. Natural erosion occurs, and without vegetation to stabilize the surface, landslides and rockslides will happen. Most importantly, mining impacts to surface and groundwater. As wells are pumped in an area, contaminants travel through the underground system, flowing to wells, housing developments, or ending up in surface waters such as rivers, streams, and lakes.

In addition, secondary impacts are indirect and often unintentional, however, they are complicated. Underneath a Louisiana bayou lays the Napoleonville salt dome which extends thirty thousand feet beneath the surface of the Earth, with huge salt pillars upward from the main salt dome. In 1982, a company installed a well to extract salt hollowing out a huge cavern. The cavern is now thought to be the culprit for the Bayou Corne Sinkhole, which is 325 feet across and has decimated the local community and continues to spew forth flammable methane gas. (Ditthany, 2018).

What needs to happen is that old sites need remediation. Just planting vegetation in barren abandoned open pit mining sites can eliminate or cut mining dust. Better management of superfund sites. Often these sites have test wells, and instrumentation. Now that we have better technology these sites need to upgrade. Most of all it is a community of concerned citizens that have the biggest impact. In Palmerton, cleanup of the town, Blue Mountain, and Cinder Bank, with oversight by the EPA began in 1987. It has included activities such as grading, revegetation, cleaning of residences, cleanup of surface water, and water treatment. NOAA and other federal and state agencies, comprising of the Natural Resource Trustee Council reached a settlement for damages to natural resources in 2009. Over 20 million dollars has been paid to compensate the Commonwealth of Pennsylvania for damages to the Leigh River watershed (NOAA, 2020).

In August of 2013, a group of local concerned citizens, the Palmerton Natural Resource Trustee Council and their partners announced the acquisition of more than three hundred acres for state game lands and the Cherry Valley National Wildlife Refuge, which is home to the endangered bog turtle just thirty miles from Palmerton (NOAA, 2020).

This group has designated other properties for restoration including habitats along the Aquashicola Creek and its tributaries. By acquiring and protecting these lands and waters this local group of concerned citizens is making up for the loss of both natural areas due to mining practices. According to the Chamber of Commerce of Palmerton the area is growing and making progress to leave their complicated mining history in the past (NOAA, 2020).

REFERENCES

CDC. (2019). *Potential for Human Exposure.* CDC. https://www.atsdr.cdc.gov/toxprofiles/tp60-c6.pdf

Diggings. (2020). *Friedensville Zinc Mines.* Diggings. https://thediggings.com/mines/usgs10067405

Ditthany, M. R. (2018). *The Pros and Cons of Open Pit Mining.* Sciencing. https://sciencing.com/open-pit-mining-pros-cons-12083240.html

EPA. (2020). *National Priorities List and Superfund Alternative Approach Sites.* EPA. www.epa.gov

Hall, P. (2021). *Pennsylvania Officials reach a $7.6 million settlement with American Zinc Recycling in Lawsuit over Pollution from Palmerton Plant.* The Morning Call. https://www.mcall.com/news/pennsylvania/mc-nws-pa-palmerton-american-zinc-recycling-settlement-20210210-fsvvg25d3vfr7kl6fg7atlekvy-story.html?outputType=amp

Kaas, M. L. (2018). *19th Century Zinc Mining in the Friedensville, PA. District and the birth of the U.S. Zinc Industry.* Lehigh University. https://www.lehigh.edu/~gpl0/presentations/KaasMikeFinal-SIA%202018.pdf

NOAA. (2020). *A Pennsylvania Town Moves Beyond Toxic History of Denuded Mountains and Contaminated Creeks.* The Office of Response and Restoration, NOAA. www.response.restoration.noaa.gov, web.

USGS. (2020). *Water Resources and Mining.* USGS. www.usgs.gov

Chapter 5
The Potato Fields and the Spring

ABSTRACT

Occasionally you meet some of the most interesting people. Often, they are neighbors with completely different attitudes and habits. Among the best locations the author discovered was one of the most prolific and beautiful spring sites. It was sandwiched between two large potato fields. Off the side of one of these fields sat a pond with a small run-down cabin on its shore. It was owned by a man of dubious character. Being one of the three adjoining owners, his only interest was to make quick, easy money. The owner of the potato fields was ill and wanted to sell his land to help his family and leave a legacy. The third owner, who actually owned the spring, was honorable (grand dame), who thought on a higher level than the other two. She was very intelligent, and the author grew to admire her and today misses her dearly. She was an example of how to be a lady and still be smart and cunning. Through autoethnography, this chapter explores those potato fields and the spring.

Thousands have lived with love, not one without water. – H. Auden

DOI: 10.4018/978-1-7998-9691-3.ch005

MEMOIR

After renting an infrared camera from NASA, I was flying over some properties looking for spring water sources, I made grids of potential areas. The ideal time to do this is in August because the ground is warm, but the spring water is cold. I sectioned off one-mile parcels of land into grids, and the entire month of August flew over these grids, taking GPS readings where a potential site might be found. It really was great fun. The pilot strapped the camera to the side of his helicopter, and I was strapped in as well. The camera and I hung a little out of the helicopter so it was an awesome ride. What a view! On one fly over, in the middle of a field, we got a drastic temperature change. It was a little oasis in the middle of the field. I suspected a sizable spring. At the end of the day, I researched the area and found there was a fishing club in the oasis. I went to the town office and found the property owner of my oasis and the farmer that owned the fields. In addition, I found a property owner for a very tiny cabin on a quarter-acre lot on a pond by one of the potato fields.

The potato field was a perfect place to build a plant, and the spring was upgradient which meant it could be protected if we did build a plant. It would take effort, testing, and loads of fieldwork to see if it could be feasible. I contacted the farmer first. He seemed excited and said he was contemplating selling his property and had already contacted a sub-division developer. He liked the idea of the plant, which would hold a LEAD (Leadership in Energy and Environmental Design) rating by the American Green Building (USGBC) More on this later! The town would rather have a manufacturing plant than a subdivision; the town needed jobs. The farmer was a character, old school, probably not completely trustworthy, but a nice guy. He was not well educated but one of those older, experienced by life, and what a friend referred to as an old brook trout. Meaning he was smart and thrived by being savvy and working hard. (Meandering through life in the stream, never getting caught, and growing his business) You just had to like this guy!

The farmer had children, but none wanted to farm. His wife drove a new Mercedes SUV, and he lived a good life. Every time I saw him, he would reminisce about his first wife and a 10-year-old boy who died tragically in a snow fort. Since I came from a farm family, I knew how hard he worked, day and night. His hard work is why his wife had nice things and he lived a good life. In his fields and on his tractor was his happy place. Cancer was ravaging his body, and he knew it was not long before he could no longer do what he loved. It was time to sell, take care of his family, and rest.

The spring owner met her husband in World War II, she was a lovely French girl and he was an American pilot. He was dashing and handsome, and she was sophisticated and beautiful. They married in France and soon spent duty time in Morocco before coming to the U.S. Her husband flew B17s and B24s in the war. If you know anything about WWII, you know that the B17s and B 24's were prime aircraft of the day, and Morocco was known as the world's spy capital. This man was a decorated soldier. I never asked, because I also come from a military family, and just do not ask those questions. She showed me pictures of them in their youth, and they were impressive. A dynamic global couple.

She shared stories about the war, and I told her she should write a book. Her grandfather lost his house in the Great War, and again in the second World War. He was a cellist in the Paris Symphony. She talked about being a young girl living in a war zone. Coming to the United States had been such a blessing and surprise. How different it was to live in freedom and the many advantages we have as American citizens.

Her husband was originally from a French-Canadian community, the same one that my husband grew up in, and we knew some of the same people. We had an instant bond and friendship. I grew to feel as if she was another mother. They found the property which at the time was a fish hatchery and purchased it. They kept the hatchery and developed a popular fishing club. In addition, this was a historical property. The king's highway went across this property and two of the trees on the property had the mark of the King of England on them. On this property were the most beautiful springs. The spring-lined streams down to the hatchery. The water was 30 degrees year-round. Total dissolved solids around 35, a perfect glacial spring made by the glaciers melting, which started around thirteen thousand years previous.

Once we received permission, we scheduled a day to perform field work and water testing. The owner of the spring did not want us to disturb her fishing guest, so we agreed to park on the other side of one potato field and walk in. A consulting company that I hired and worked with for many years always had one of the staff accompany me. One young geologist who was like a son went with me often. He was a kind, generous big guy, rugged and smart. He was designated to go with me that day. In the previous chapter, I went into detail about how affected I was by dog poop. I said there was something much worse. I was about to experience that beautiful summer day. The skies were blue, it was 75 degrees when we started out. As we got into the field, the foulest order and the overwhelming stench hit us. Both of us started sweating, and of course, I started vomiting. The farmer fertilized his potato fields with fish parts. He got this great fertilizer free from a fish processing plant on the coast. Fish fertilizer is made from fish carcasses, bones, scales, and skin. It was decaying in the summer sun. Oh, the joy of it!

We were dying, and I am not saying this likely. My only hope for survival was to get off the field. When we did, we hit lush green grass and could hear water running in the distance. At that point I did not even care about the spring, I just wanted to get as far away as I could from the potato field, and I was not alone. My young companion was as bad off as I was. When we saw the stream, we just literally laid down in it. We did not care what was in it, we did not care how cold it was, we just immediately laid down, boots, clothes, and all. We laid there for about ten minutes, not noticing our environment before we could move. I sat up and saw the most spectacular site. It looked like a rrainforest A scarlet tanager was on a tree branch right in front of me. I looked over at my young companion and said, "I do not care if I have to walk fifteen miles back to the car, I am not crossing that potato field again". He quickly agreed, or he would have been waiting for me at the car for a long time.

Once we both got our bearings, we realized how precious this site was. How beautiful and pristine it was. When finding something like this, which is rare, I knew the great responsibility I had to save it and protect it. We took our water samples, took field notes of the area, and started to find a way out. We had our compass and maps, so we knew we could cross the property to a housing development five miles away. Walk the highway and back in. I figured it was around a fifteen-mile hike.

As always with nature, we had one more surprise. Almost at the end of the property was another spring-fed spring, and we came across an area that looked like a bomb had gone off. It took me a minute to figure out what had happened and what it was. My first thought was a contaminated site, but then I looked at the gnawed-off stumps of small trees and knew. Who knew that sweet gentle creatures would create this all the while building a wetland? Beavers! When you have beavers on your property, it is a good sign that this is a special place.

We walked and walked, and finally found the subdivision and circled it to the highway. Then we came up a private road to the potato field, a long walk back to our cars. It was late in the afternoon, and I looked at the fields before leaving. The one on the other side of the field we walked into was a perfect plant site. It was a large field with a road in-between, this led to highways. On the other side of the hatchery, is another large potato field. I would have to contact the owner of the small run-down little cabin. The farmer warned me that he might want to take advantage, and he was not a pleasant person to deal with.

The samples were sent out to seven different laboratories. While I was waiting for the results, a surveyor friend and I did research in the town office on land use. We were able to go back to the 1800s but not before. We were told by the town assessor that if we wanted to go back further, we would have to go to the federal courthouse just south of Boston, Massachusetts. Maine was once a part of the Commonwealth of Massachusetts. So, the next day we took the train to Boston and the subway to the federal courthouse. A treasure trove of historical information. Both of us love history, so we spent the entire day. I found historical information about the property going back to the early 11600s. The part of the property and the surrounding area was sold in 1619 to Lieutenant William Philips of the English army by a native chief. The chief's mark was a feather. Then later the rest of the property and more surrounding area in 1625 was sold to Captain William Philips, the same as above, by a native chief, and his mark was a bear. I learned that this was the mark for the famous Chief Sunday.

The property for the rest of the 1600s was cropland farms, mainly vegetables. The existing field on the other side of where I staked as a plant site, was called the California field. In 1725 the native Indians revolted against the white settlers and started massacring families. Chief Sunday was able to warn many families and led them to safety. Fortunately, the property that was the potato fields were farm fields, except for a small piece that was a cemetery.

When you buy property in Maine, and you find historical evidence of a cemetery, you must locate it and preserve it. I walked the property and could not delineate the cemetery. I talked to the farmer, and of course, he said he had never seen one. Hmm! I contacted the Department of Environmental Protection in Maine, to arrange for one of their staff to be on-site, and hired a company to GPS the entire site, in hopes of finding the cemetery. Then we could fence it off and mark it properly. This took place over two days, and we found no evidence of a cemetery or bones. The Department of Environmental Protection (DEP) and I agreed that our trusty farmer most likely plowed it up and got rid of it. He had a trash pile, but over the years it was hauled off. The DEP agreed that there was nothing I could do, we were not building on this side, but just wanted to keep it from any development. We agreed.

We purchased all the property and the spring, but the cabin owner held out because he had access to the cabin with a road in one of the fields. He wanted what we paid for everything. I said no. We started our permitting process. I started the process of getting to know my neighbors and teaching geology to local citizens at the town office. People get nervous if they do not know you or what you are about. It is best, to be honest, and up-front. I found over the years that adults either forget what they learned in school or were not taught earth science. I found over the years that all adults like learning about science and the history of their own community.

After three months we had town planning board meetings. This is when the cabin guy showed up with one of his buddies that were a Harvard professor in economics. On another spring development, we had an issue with a spring owner who was a Harvard alumnus, was not honest with us about his spring, and we quit using it. From what we understand, he joined forces with the cabin owner to cause trouble for us. He continued to show up through other people because we had restraining orders against him. Remember, the Unabomber was a Harvard graduate. I found out later he was coaching our cabin owner and got some of his acquaintances to show up at our planning board meeting and was paying them. When we went through the science, they wilted and left. Speak the truth in all things and at all times, and it will serve you well.

THE PROJECT

Once we received town approval it was on the various state agencies. My part of this $120 million project was $45 million. It was to develop the Spring houses, pipeline to the plant, storage tanks, long-term monitoring, and modeling programs. In addition, I was to do the DEP permitting for the entire facility. This included the LEED work on light pollution, stormwater catchments, and grounds. Two engineers build the plant, production lines, and everything else.

U.S. GREEN BUILDING CERTIFICATION

The (USGBC) is a leader in helping communities and cities safely, healthily, and sustainably thrive. This is not about construction but about creating the world we all want to work, live, and play in. "*LEAD is for all building types and all building phases including new construction, interior fit-outs, operations, and maintenance core and shell*" (UAGBC, 2021). LEAD has five main ratings:

- Green Building Design and Construction

- Green Interior Design and Construction
- Green Building Operations and Maintenance
- Green Neighborhood Development
- Green Home Design and Construction

LEAD has a point system to score all the green projects that apply to the process. This point process rates projects with certified silver, gold, or platinum. Seven impact categories are identified. The overall themes are climate change, human health, water resources, biodiversity, material resources, and a greener economy and community (UAGBC, 2021). At the time this was revolutionary, and to start this process you need help. The is an exam for engineers and provides certification for LEAD and green building. We started this process as novices. We were building this project and another one in another state. Our end goal was to have all our facilities with a LEAD rating.

ROCKY MOUNTAIN INSTITUTE

We hired one of the best organizations available, and they still practice their craft and are leaders in the environmental movement. This is the Rocky Mountain Institute (RMI) in Aspen, Colorado. RMI is a non-partisan, non-profit, independent organization of experts of many disciplines who are working to accelerate clean energy and improve the quality of life for all life on the planet. RMI was founded by Amory Lovins, who is now Chairman Emeritus. Today, RMI has over 300 staff working around the world on projects on a broad range of global issues.

THE PROJECT

Jobs were needed in this area, and our plants are clean and well maintained. Because we are a natural resource, the plant is regulated by the Department of Environment Protection and the EPA. Because we make packaged food products, we are regulated by the state health department and the FDA. In some areas of the country, we are regulated by the Bureau of Land Management. The permitting and testing to build one of these plants is $500,000.

Construction to build a green industrial facility raises the construction cost by 25%. This was a $120 million project. The entire spring lines, storage tanks, and production lines are all 316 low carbon stainless steel. This steel is of the highest quality and used for food processing. Most stainless steel ordered around the world is Grade 304. It offers the standard corrosion resistance, formability, strength, and easy maintenance for which stainless is known. While 316 comes in second in terms of quantities sold, it offers vastly superior corrosion resistance to chlorides and acids.

Each plant has a Hazard Analysis Critical Control Plan (HACCP). This plan is a food safety monitoring system that is used to identify and control biological, chemical, and physical hazards within the storage, transportation, use, preparation, and sale of perishable goods. It also determines critical control points (CCP) in the process of food production. There are seven steps in the HACCP plan:

- Perform a hazard analysis.
- Determine Critical Control Points (CCPs)
- Set critical limits.
- Establish a monitoring system.
- Establish corrective actions.
- Establish verification procedures.
- Establish record-keeping procedures.

These plans are usually put together as a joint effort of different departments and housed in Quality Control. Different groups handle different facets of the beginning permitting, monthly and yearly monitoring and permitting. My part is for construction and source monitoring and permitting.

DEVELOPING PRISTINE SUSTAINABLE SOURCES

It all starts with geology! The geology of this state and the area goes back to the glaciers that sat on the land from the last ice age. The Pleistocene Epoch is typically defined as the time period that began about two million six hundred thousand years ago and lasted until about eleven thousand seven hundred years ago. The most recent Ice Age occurred then, as glaciers covered huge parts of the planet. Northern parts of the United States were covered in glaciers miles thick. There have been at least five documented major ice ages during the four billion six million years since the planet was formed. There may have been more.

The Pleistocene Epoch is considered with a man as we know it evolved and could be found throughout the earth. The Pleistocene Epoch is the first epoch in the Quaternary Period and the sixth in the Cenozoic Era. The current epoch the Holocene Epoch is the current stage we are in which followed the Pleistocene Epoch. At the Pleistocene time the continents had moved to their current position (Zimmerman, 2017).

Within the glaciers were movements over time, around twenty cycles when the glaciers would advance and retreat – thawed and refroze. The climate was much colder and drier than it is today. There was so much ice on the planet there was little precipitation; half of what it is today. During peak periods with most of the water frozen, global average temperatures were nine to eighteen degrees F. There were winters and summers during this period, but variations in temperatures produced glacial advances because the cooler summers did not completely melt the snow. While mammals emerged, and man evolved, many species such as large mammals did not, such as the woolly mammoth, saber-toothed cats, and giant sloths. Many species did survive, such as man, birds, and reptiles. Vegetation was limited in many areas, with scattered conifers, along with broadleaf trees. Prairie grasses and flowers such as lilies, orchids, and roses were abundant. Around thirteen thousand years ago, more than three-fourths of the larger Ice Age animals died out. Many scientists still debate and do not know what caused this extinction (Zimmerman, 2017).

When the great glaciers melted, they often melted fast and from beneath. As they melted the fine sands and gravel fell out of the glaciers. The larger material melted out last. This fine sand and gravel were filled with melting glacial water forming the prolific sand and gravel aquifer systems in the northern part of the U.S. Porosity determines the capacity of the material to hold water. Sand and gravel are not as porous as other materials, such as clay, but they are more permeable. Permeability determines its ability to yield water. For example, clay is made of tiny particles with a large amount of pore space between them. However, the pore spaces are so small that they create a resistance to flow which reduces groundwater permeability. Permeability is an important characteristic since it determines whether groundwater can be drawn into a pumping well (MGS, 2020). Sand and Gravel aquifers hold an enormous amount of water, they recharge easily and make an ideal water source. In addition, they have lower dissolved solids, and water in known to have a great flavor. They are natural sand filters. This source was a sand and gravel aquifer.

Our work started with the investigation, site walk, and testing. Once we found a potential site, we flew over it in a helicopter, I have spent many hours in a helicopter. Now resource managers use drones. Next is an actual site walk, making sure we investigate anything of concern or interest on the ground. As we did this intense site walk, we took water samples. Some of these would be performed in our plant laboratories, and some would be sent away for detailed analysis of specific chemicals and minerals, testing we could not perform in our own laboratories. While this was being done, an extensive investigation was being performed in courthouse and town records. This has been stated before, as well as making sure to go back one hundred years to provide the most detailed study of land use. This was only the beginning!

On this property, all the above was being done before negotiating on both properties. Sometimes this is a long and difficult process. Land owned generationally by families is often sacred. This I understand from my own family. Once the land has been sold it is often gone forever. When the property was purchased, the real work began. I had the finest surveyor, consultants, and engineers a project manager could have. The expertise of this group was phenomenal.

We as a team started right away. Time was critical. We had weekly team meetings and monthly team meetings with the overall project team (the team building the plant). A detailed survey was completed as part of the purchase agreement. Using that survey, we developed a preliminary testing and analytical plan to map the aquifer and model the aquifer. Small 4-inch wells were drilled in specific areas, where we could measure pressures and water depths. These are called piezometers. In addition, we installed shallow monitoring wells. Both allow groundwater to rise and fall and monthly measurements were taken, fed into a computer software program so that in-depth analysis could be done over time. When piezometers are placed apart from each other, they can be used to estimate the local depth of the water table in relation to rain and snowfall. In addition, if the elevation of groundwater at one well is lower that the elevation of another well, two are hydraulically connected, then it is assumed that the water is flowing from the higher source to the lower source. The difference in elevation between the two sites is known as *hydraulic head*. Putting dye into the upper well and monitoring to see if it shows up in the lower well can indicate *water velocity* (Dodd & Whiles, 2010). *See chapter fifteen for definitions*. Besides taking readings of well levels, water samples are also taken monthly. This data is used, along with additional data to perform groundwater modeling.

Staff or stream gauges were put in the streams to monitor water movement at any given time. "Stage", is the water level above some arbitrary point in the river/stream and is commonly measured in feet. On any normal day, when no rain has fallen for a while, a river/stream might have a stage of two feet. After a big thunderstorm the stage gauge could rise to fifteen to twenty feet, and sometimes quickly. Record keeping is important because it will let us know when water overflows the banks it is flowing through, or when wells being pumped need to be slowed, shut down, or have a danger with an influx of water that may affect water quality (USGS, 2020). More on this in a future chapter.

To monitor weather and climate I installed weather stations that were tied into NOAA's database, and fed into the National Weather Service Center in Boulder, CO. Weather is the short-term state of the atmosphere in the past, present and future. The conditions we experience include temperature, pressure, humidity, precipitation, cloudiness, fog, floods, storms, wind, and other variables. Climate, on the other hand, are long-term changes and observations of the atmospheric conditions. Weather may be cloudy and wet in the morning, and sunny in the afternoon. It may be snowing in Maine, but hot and sunny in Arkansas. Descriptions of weather include both time and location (NOAA, 2020).

Climate is the average of the weather monitored over the long-term. The expression of long-term usually refers to thirty years or more. This timeframe establishes the range of weather conditions at a given location throughout the year. When looking at weather over decades you may be able to summarize them by saying summers are hot and humid in Arkansas, but cold and windy in Wyoming. This provides a sense of what to expect in each season, they do not prove details about weather on any given day (NOAA, 2020).

Climate modeling helps scientist understand complex systems and find possible solutions. We use models to understand how things work. Climate models are known as general circulation models that use mathematical equations to characterize how energy and matter interact in different parts of the ocean, atmosphere, and land. This is a complex process that sets variables to represent conditions and changes in climate by repeatedly solving the equations using supercomputers (NOAA, 2020).

Once a climate model is developed, it will be tested via a process known as "hind-casting." "*This process runs the model from the present time backwards into the past. The model results are then compared with observed climate and weather conditions to see how well they match. This testing allows scientists to check the accuracy of the models and, if needed, revise its equations. Science teams around the world test and compare their model outputs to observations and results from other models*", (NOAA, 2020).

Groundwater modeling is not that much different than climate modeling. It is the groundwater flow that is represented by mathematical equations of groundwater flow through an aquifer (see definition in chapter fifteen).

Groundwater flow models are as complex as the real system beneath the ground. They integrate and assure consistency among aquifer properties, recharge rates, discharge rates, and groundwater levels. Models can be used to estimate flows and aquifer characteristics, stimulate the response of the aquifer to hypothetical conditions, and identify sensitive areas where additional hydrologic information could improve understanding of the aquifer (USGS. 2020).

This is just the monitoring; construction is another detailed process to the development and management of a site. As stated previously, all borehole development is done with three sixteen low carbon stainless steel pipes, pumps, screens, pipelines, and equipment in spring houses. Once the borehole is developed, this is a well that correlates to a spring. You must show the well has the same chemistry and geophysical properties as the spring. By pumping the well, it shows a connection to the spring, but you cannot deplete the spring. You can only take what nature has given the spring in precipitation each year (FDA, 2020).

A spring house is constructed around the borehole. This captures information about the well and water electronically. A quality control technician comes daily to retrieve water samples to test PH, bacteria, temperature, and any other additional information that seems necessary. Monthly, a series of water tests are sent to outside laboratories for testing, and yearly a complete battery of tests is sent out to laboratories for water testing. This entire process cost around $250 thousand yearly. This does not include the permitting. Permits are with state environment offices, state health offices, and the FDA. Regular data is collected daily, monthly, and yearly to keep permits current. Most facilities' natural resources departments work with quality control to ensure that HACCP programs and all other permits are complying.

Internally, the complex modeling and monitoring programs are managed by natural resources, and analysis is performed weekly, monthly, and yearly to ensure a sustainable aquifer is maintained. Without this development and more like it, the local community and the state would have no information on the aquifer and the area's land uses. Often these projects are criticized. Objections are based on emotion and not science. When making decisions that affect your local community, could help provide jobs, people need to make educated informed decisions.

During work on this project, it was important for our neighbors to know us, understand we utilized the science, and trust us. In surrounding towns, many educational and informational meetings took place with planning boards and citizens alike. As a developer, it was our responsibility to develop relationships with all residents of the town and surrounding areas. The local community and state government officials, needed to understand the science and trust our consultants and us to know that we had every intention of maintaining a sustainable resource for future generations. A responsibility I and my team personally took very seriously.

REFERENCES

EPA. (2020). *Water Resource Management*. EPA. https://www.epa.gov/ground-water-and-drinking-water/resource-management-protecting-your-drinking-water

FDA. (2020). *Bottled Water*. FDA. https://www.registrarcorp.com/fda-food/state-bottled-water-requirements/?lead_source=Google%20Ad&utm_source=google&utm_medium=cpc&utm_term=fda%20bottled%20water%20regulations&utm_content=109908280218&utm_campaign=11069551891&matchtype=e&device=c&gclid=CjwKCAjwvMqDBhB8EiwA2iSmPOX-8PXeTw5lPsK6j_qsAB3TIyHGwzn_olB-JMRkTQPX7TgTTPGCTBoCuFwQAvD_BwE

MGS. (2020). *Sand and Gravel Aquifers*. Mane Geologic Survey. https://www.maine.gov/dacf/mgs/explore/water/facts/aquifer.htm

NOAA. (2020a). *Weather and Climate. National*. NOAA. https://www.ncei.noaa.gov/news/weather-vs-climate

NOAA. (2020b). *Climate Modeling*. NOAA. https://www.climate.gov/maps-data/primer/climate-models

RMI. (2020). *Sustainable Development*. Rocky Mountain Institute. https://rmi.org/

USGB. (2020). *LEED*. American Green Building (USGBC). https://greenbusinessbureau.com/?gclid=CjwKCAjwvMqDBhB8EiwA2iSmPJ42zPIj5oAaAlfvvygDY_bAsLf6Y2jG66VNCiOXYIuLWoDneuE12BoC2hUQAvD_BwE

USGS. (2020a). *Groundwater modeling*. United States Geologic Survey. https://water.usgs.gov/ogw/gwrp/activities/gwmodel_dev.html

USGS. (2020b). *Piezometers and Groundwater level.* United States Geologic Survey. https://www.usgs.gov/centers/ca-water-ls/science/piezometers-and-groundwater-levels?qt-science_center_objects=0#qt-science_center_objects

USGS. (2020c). *Stream Gauging Basics.* United States Geologic Survey. https://www.usgs.gov/centers/ca-water-ls/science/piezometers-and-groundwater-levels?qt-science_center_objects=0#qt-science_center_objects

Zimmerman, K. A. (2017). *Pleistocene Epoch: Facts About the Last Ice Age*. Live Science. https://www.livescience.com/40311-pleistocene-epoch.html#:~:text=The%20Pleistocene%20Epoch%20is%20typically,parts%20of%20the%20planet%20Earth

Chapter 6
Meeting the Mob at Denny's and the Ancient Spring

ABSTRACT

These two observations are put together in one chapter even though they are entirely different stories. The outcome is the result of the same sustainable, quality issues, or lack thereof. People go about their daily lives just trying to pay their bills and build good lives. They often do not truly understand the consequences of their actions unless they are educated in a specific area. The author does not like politicians acting like experts on science or climate change or environmental issues because they often get it wrong due to a lack of knowledge. It is best to trust the scientists to tell us what we need to know and educate our adult community on the basics of sustainability.

When life places stones in your path, be the water. A persistent drop of water will wear away even the hardest stone. – Autumn Morning Star

MEMOIR

Early on in my career in natural resources, we were building a new plant. This time my job was to find all the spring sources for this new facility. Thus, my trip to the zinc mines. I received a call from a man who said his family

DOI: 10.4018/978-1-7998-9691-3.ch006

once bottled whiskey, and they used a beautiful spring that was the source for their water. He touted the spring water enhanced the taste of the whiskey. He even sent me pictures of the spring, bottles of whiskey, and labels. I was excited. It was not close, but not so far, that we could not make it happen. So, I planned to meet him at a Denny's to go to visit the site. I had my field equipment and testing bottles packed in my field bag and took off. It was a beautiful summer day, a great day to be in the field.

I must have looked like the typical geoscientist. I had my jeans, boots, company shirt, and field vest on. I pulled up to the Denny's and started to walk in when a very tall, very big man in a very expensive suit, wearing an earpiece, asked me if my name was Kristin. I looked up, thought he was impressive looking and said yes. I thought this is very strange. Is this the guy who I am supposed to meet, and why the earpiece? It got stranger very quickly. He escorted me to the back part of Denny's where a half-moon table had his twin standing next to it, and two men sitting in the booth. One at the very back and one to one side. I was shown to the other side. The two big guys stood opposite each other. I sat down, feeling very uncomfortable.

The guy sitting across from me was introduced as Linnie, the person who called me, and whose family had the spring. He was older, maybe mid-seventies or early eighties. He wore a nylon tan jacket, a golf shirt, polyester brown knit pants, and later I would see his shoes were white loafers. His hair was fashioned in the 1950s style like Elvis. I thought Leisure Suit Larry! Just like the video game that came out in the 1980s. The guy in the back, then sort of introduced himself. He did not give a name but said he owned several restaurants in New Jersey. He wanted to get into the water business and wanted to be in business with us. He was an extremely handsome gentleman. I guessed his suit was the most expensive suit I had ever seen, and I thought to myself, "now I know what they mean by a man's expensive suit". He had a gold and diamond ring on each hand, one was a pinky. He smelled wonderful, polished, and well-spoken. The meeting did not last long. He said he had other business. He looked over at Linnie and told him not to mess this up and we left. He shook my hand, and held it slightly longer than I thought appropriate, I thought. "this is a very dangerous man, one who knew how to charm anyone, and I was really confused. Is this the mob?"

I was to ride with Linnie to the spring. So, I grabbed my field bag out of my car and got into Linnie's big ole silver Lincoln. As I sat down, I thought "I am so naïve, just a little farm girl from Kansas". I prayed a silent prayer that God would watch over me, "I said God, I have a husband and five children at home that need me, so keep me safe". On the way to the spring, we passed corn field after corn field. The thought came to me, "I could be buried out here and nobody would find me." My kids and husband would never know what happened to me. The thought hit me; I did not tell anyone at the plant where I was going. Oh my God!

We arrived at an area where a big white house stood over a huge lawn, down to a fenced-off area of about half of an acre. In this area were a cow, a pig, and hay. He told me his nephew lived in the house, and these were his animals. The spring was in the middle of this half-acre parcel. I asked him how much land they still owned and he said this was it, that most had been sold, and the old liquor bottling plant had been destroyed. Hmmm!

This would not do, but I did not say anything. We went through the gate into the area and to the spring. There was a round concrete lid, that was very hard to lift. It took both of us to lift the lid. Inside was water, not flowing, and debris. I collected my water samples, and we headed back to Denny's and my car, as we pulled up to my car, I could feel the relief through my shoulders. I got out and told him I would test the samples and let him know the results. Linnie walked me to my car, as I started to get in, he pulled the largest wad of money out of his pocket and asked me if I needed help. I am sure my mouth dropped, but I said no that my company took very good care of me and made sure I had what I needed to do my job. I got in my car, and when Linnie walked off, I locked my door and started my car. Just so happy to be going back to the plant.

All the way back, I just kept thinking of my husband and kids. I realized for the first time in my life that I was very naïve. How could I be this naïve? I lived and worked internationally. I had lived in several parts of the country but had never been exposed to the mob. It also occurred to me that this is just another example of people ruining their property by not knowing anything about natural resources. I knew the samples would have fecal coliform in them by just looking at the land use. The tests showed much more than fecal coliform. The samples showed high concentrations of contaminants from the farms surrounding the area. I called Linnie, gave him the results, and suggested he contact his nephew in the house to install a water purification system on his incoming water well.

Upon returning to the plant from meeting the mob, I went to the plant manager's office, a good friend. I do not know what hit me, but as I relayed the story to him, I got teary-eyed. He called the corporate office and requested field help for me. I was given a summer intern to ride along with me. You might think they might hire a young male geology student. Oh no! They gave me the most delightful twenty-year-old girl studying environmental science.

She was the daughter of a woman who once opposed our plant and searched for springs. She ended up being like a daughter to me and the perfect fit.

I had met her mother at a town meeting. Although she later became a friend, she was the opposition; her friends had come ready to have it out with us. Once I presented the science; I could tell she was changing her mind. One of her friends was not listening and yelled at me that my children were ashamed of me. I stopped and said in a nice even tone. No, they are not, all five of them love me, and they are very proud of their mom and how hard I work for them. The room fell silent. It was a turning point with the naysayers because they started to listen. When they really listened, they started to understand the science and what the new plant could do for their community. After the meeting, I went and asked several of them for coffee.

We became friends and had a great time. Your enemy is often your friend if you just give it a chance. It was during this time we were looking for an intern. My friend called one day to ask if we had a summer job for her daughter. I hired her to be my intern. We spent the summer looking for spring water in very unusual places and grew very close.

FIRSTHAND OBSERVATIONS

The provinces of Canada were in my territory! We had beautiful plants in which I was tasked to find more spring water for them. One day I received a call from a spring owner, and he sent me some information about his spring. It literally was a documented ancient spring. The geology of the area was clear. This spring had probably been there since the melting of the last ice age. In the middle of the 1800s, it was encased in marble and a beautiful spring catchment was built around it. Records showed it had a consistent flow year-round. The spring owner had inherited the property and was looking for revenue to maintain it. In addition, it had a beautiful home a mile away and was surrounded by land. Some of the lands had been sold to an area farmer, but the owner controlled a thousand acres and the recharge area. The best for us was, it was only forty-five miles away from our plant.

When I arrived on site, I met the owners, a lovely couple, full of passion for their property. I was so hopeful that we could work with them. The buildings on the property were beautiful, the entire complex looked like an ancient monastery. The energy was one of peace and relaxation. I took my samples and then was given a tour of the property. I did notice several elk herds on the property and was told they were generating revenue for the property. The antlers were sold to Eastern medicine outlets and the meat to high-end restaurants. The hides were sold to a leather works factory. Nothing was wasted.

I did not think anything of it, being a girl from Kansas farmland and ranchland, but I did ask if they rotated the herds. The person taking me on the tour did not know and I recall thinking at the time," this is a large herd of elk". Sometimes, our intuition is screaming at us and we just do not listen. I went to Fed Ex and sent my samples and boarded my plane home. Two weeks later, the lab results came back and I was surprised to see fecal coliform and other contaminants. Finally, the light bulb turned on, the elk! They were overgrazing and had too many animals on too small of an area of land. This was causing the groundwater to be contaminated.

I called the spring owner, and he was devastated. This was a shock. I told him I would get expert advice from his own country's agriculture agency and see what could be done. Until this was solved, I could not use the spring. I also informed him that this could take years of non-use to remediate the issue. Of all the people I met, I have no doubt they would do the right thing. They are still there, caring for the property, and still have the passion to develop it for a great cause.

LAND USE

Sustainable land use should not be so hard, but people do not understand much about it. They do crazy things with their land because they lack a basic knowledge of how to protect land and water resources. The term "land use" describes how people use land. It is a term that defines the economic, agricultural, residential industrial, and recreational use of land. Not just private land but public land as well. For sustainability of natural resources, it is necessary that land management be practiced.

Good land practices include protection for soil, topography, water, vegetation, and animal life, including human life. This can be difficult when addressing private property rights. Private owners may feel they have the right to manage their land any way they want. Unfortunately, this may affect their neighbors, or water downstream from their property. Most governmental agencies recognize five types of land use: residential, agricultural, recreation, transportation, and commercial. Land resources are used for food, shelter, and economic development. Sustainable management of land resources is crucial to protecting ecosystems, watersheds, and biodiversity systems (USDA 2020).

Land use information details how our land resources are to be managed. Land mapping is based on remote sensing, which gathers data from satellites, drones, or helicopters. This is important information used to define and determine land management practice and use. Cropland, forestland, pastureland, and rangeland comprise most of the land uses in the U.S. and the world. It is important that this land uses practice sustainability for soil, water, air, and animal habitat. The role of governmental agencies is to provide education, leadership, and technical assistance.

The USDA has created the Natural Resources Inventory (NRI) which collects and produces scientific information on the status, condition, and trends of land, soil, water, and related resources on all public lands and waterways. The Land Evaluation and Site Assessment (LEA) aids state and local governments to make sound decisions about land use. It provides a framework to numerically rank land parcels based on local resource evaluation and site considerations (USDA, 2020).

Agricultural land evaluation starts with soils and how they are related to other groups, such as cropland, forestland, or rangeland. The value is determined for each group and the total value is based on data from the National Cooperative Soil Survey. This database is called the most valuable natural resource database in the world. Sadly, in 1980 it revealed that millions of acres of farmland were being converted in the U.S. each year. Programs were implemented to protect farmland and combat urban sprawl. The data showed that much of this urban sprawl was the result of programs funded by the Federal Government. Congress passed the "Agriculture and Food Act of 1981, containing the Farmland Protection Policy Act. On June 17, 1994, the final rules and regulations were published in the Federal Register (USDA, 2020).

Agriculture is one of the largest users of surface and groundwater in the U.S. Agriculture irrigation systems use the most. In 2000, almost thirty-four percent of the water withdrawn from surface water and groundwater was used in irrigation. In arid and semi-arid areas, cropland is dependent on irrigation. This competes with human use of water, especially in areas where water sources have been experiencing shortages.

Agricultural runoff can impact water quality. Runoff carries pollutants to streams, and lakes and can seep into the groundwater. State and First Nations have been monitoring for nutrients and contaminants that are affecting surface and groundwater supplies. Excessive amounts of nitrogen and phosphorus will prohibit life in aquatic environments. Excessive sedimentation can overwhelm aquatic ecosystems such as freshwater breeding areas and degrade coastal and marine ecosystems. In addition, pharmaceuticals and fecal coliform from livestock and poultry are found in our water resources (USDA, 2020).

The USDA (2020) has over one thousand three hundred active or completed watershed projects. They aid and install conservation practices and project measures throughout the watershed project area. These projects are planned to take place over multiple years. They include projects such as retarding dams, and improvements to reservoirs; some are owned by local organizations or individuals (USDA).

The National Water Management Center (NWMC) is in Little Rock, Arkansas, and is the information exchange for the USDA. This agency is the support center for addressing water resource problems across the nation by providing leadership, assistance, information, and technology for natural resource sustainability. They provide expertise in and guidance for the technologies to assess watershed health and plan watershed-scale solutions. The NWMC team of scientists work with state and national specialists to develop solutions to water resources (NWMC, 2020).

LIVESTOCK

Twenty-six percent of the planet's arable land is used for livestock grazing and thirty-three percent of croplands are used for livestock feed production. It is thought that seven percent of greenhouse gasses come from fermentation and manure from livestock. Livestock provides twenty-five percent of the protein intake globally. And fifteen percent of dietary energy. The demand in animal products both meat and dairy products has risen, mostly for eggs (FAO, 2019).

Livestock is a driver of land-use change. Globally, thirteen billion hectares of forest areas are lost due to land conversions for agricultural uses for pastures and cropland for livestock feed. This not only impacts forests, but water supplies, soil conditions, and biodiversities. Removing that large amount of forest impacts carbon dioxide levels. In addition, twenty percent of grasslands are turned into agriculture and this is increasing, due to animal density per area (FAO, 2019).

Even more of a concern is the increased risk of disease from livestock in close confines with wildlife and humans. Sixty-six percent of the emerging diseases in humans have animal origins and one or two new diseases emerge yearly. Hormones and antibiotics used in industrial meat production affect human health. The FAO (2019), has suggestions to balance these issues:

- Promote and consume livestock and livestock products raised on grasslands.
- Focus on animal health.
- Eat less meat!
- Support managed grasslands!
- Reduce subsidies for Chemical fertilizers and concentrate food!
- Regulate stocking rates on grasslands and the use of antibiotics!
- Inform growers and consumers!

CARRYING CAPACITY

The carrying capacity for grazing is measured by understanding how much forage is available compared to how many animals are grazing in a certain area, or grazing capacity. This is to balance the available forage supply with livestock demand. Carrying capacity can be calculated using a variety of techniques, monitored, and adjusted over time to determine the long-term effects of grazing in a certain area. There are two recommended methods to calculate grazing capacity:

- One: gives an estimate of carrying capacity based on available forage production guidelines. This is easy and works best when the pasture conditions are similar throughout the field and the forage plant community is uniform.
- Two: can be used if they plan on clipping, drying, and weighing samples collected from their pasture. Field-based sampling provides greater accuracy but requires more hands-on work. Producers may choose field-based sampling if provincial guides are unavailable for their region or if pasture types or conditions vary within their field. Forage production varies each year, so the approach should include multiple years of sampling to estimate the long-term productivity of the pasture.

Research on the comparison of beef production and the environment in which it is produced is important in providing the knowledge and tools needed to be sustainable stewards of natural resources. Efficient and environmentally responsive use of livestock manure provides a valuable source of nutrients, organic matter, and carbon to pastures and cropland. Manure offers a long-term source of nutrients that can contaminate soil through increasing carbon and nitrogen and alter soil phosphorus and potassium concentrations along with other nutrients.

In addition, manure can affect surface and groundwater quality. The phosphorus and nitrogen in cow manure can run off with rainfall into local waterways or seep into the groundwater. In addition, pesticides, fertilizers, and antibiotics in manure do not stay stationary on the landscape where they are present but runoff, infiltration, and through precipitate, these contaminants percolate into local streams, rivers, and groundwater.

Farming is not the only source of water contamination. However, there are ways to reduce pollutants. The first is to manage runoff and leaching of nutrients through the soil and place the stockpile in an open-sided shed, on a level surface, above the season high-water table. The best is a catch basin to hold runoff before it reaches a waterway. Catch basins and grass buffer strips can be used to hold and filter runoff from open lots before it reaches a waterway. It is very important to manage leaching from storage pits to keep nutrients from leaching downward. Pits should also be monitored closely and pumped before overflowing. This can be done by using a clean-water diversion system. Berms, ditches, and gutters can be used to divert upslope and rainwater from areas with manure so that it does not carry nutrients and pathogens to waterways (Modderman, 2020).

The overgrazing of millions of sheep and goats is the primary reason the Mongolian Steppe has experienced land degradation. This area is one of the largest remaining grassland ecosystems in the world. Using satellite-based vegetation monitoring systems, research determined about twelve percent of the biomass has disappeared in the Mongolian Steppe. Currently, seventy percent (2013) of the grassland ecosystem is now considered degraded. Overgrazing accounts for around eighty percent of the vegetation loss in recent years (Hilker, 2013). According to Hilker, socioeconomic changes linked to Russia, have caused unemployment which has led many people back to domestic herding. Since the 1990's livestock has doubled. This issue means that land is being degraded and the food supply is reduced.

To prevent overgrazing, pasture forage can be supplemented with stored livestock feed, livestock can be rotated and pulled off pasture land, and a percent of pasture land can be planted for warm-or cool-season while perennial plants recover, depending on the type of plants. Basically, the lack of proper animal and wildlife feeding management causes overgrazing. Looking closer at the issue, you have to look at the socio-economic background of the area farmers. The assistance and education that is available to these farmers are free. It all comes down to land use (CBD, 2020).

The Center for Biological Diversity leads efforts to reform overgrazing on public land in the Western part of the U.S. The center's primary purpose is to protect endangered species and conserve habitats. They have determined that the federal grazing program allows beef costs to stay down. The western federal rangelands account for less than three percent of all forage fed to livestock in the U.S. If removed, beef prices may be affected. The ecological cost is high and destroyed vegetation, damaged soil, and stream banks, along with contaminated waterways from fecal waste, had eliminated forests, and many aquatic habitats. Predator control programs designed to protect the livestock industry have driven the grizzly and Mexican grey wolf out of the area CBD, 2020).

Getting back to my elk heard, by first-hand account and observations, that both parties, unknowingly, negatively affected their property. The first one did not understand how farming and the poor practice of having just a few animals can change the water quality in spring and the surrounding groundwater. I am sure the well in the large white house is contaminated. Just because the area looks quaint and they tried to maintain a "gentleman farm" what is happening beneath the ground is much different. The well going into the house needs to have a filtration system to remove the nitrates and any risk of coliform contamination. This would be expensive and would most likely take a Reverse Osmosis system or microfiltration.

A reverse osmosis system removes a large majority of contaminants from water by pushing the water under pressure through a semi-permeable membrane. Osmosis is a naturally occurring phenomenon in which a weaker saline solution will migrate to a strong saline solution. Plants absorb water from the soil, and our kidneys absorb water from our blood. A semi-permeable membrane will allow some atoms or molecules to pass but not others. Reverse osmosis is the process of osmosis in reverse. Osmosis occurs naturally without the energy required. Moreover, to reverse the process of osmosis you need the energy to move the saline solution. A reverse osmosis membrane is a semi-permeable membrane that allows the passage of water molecules but not the majority of dissolved salts, organics, bacteria, and pyrogens. By pushing the water through the membrane by applying pressure that is greater than the naturally occurring osmotic pressure in order to desalinate water in the process, allowing pure water through while holding back a majority of contaminants (Puretec, 2020).

Microfiltration is the process of physically removing suspended solids from water, usually through a membrane. This is not nanofiltration or reverse osmosis because it only utilizes physical filtration to remove particles in

Figure 1. Reverse osmosis
(Water-online, 2022).

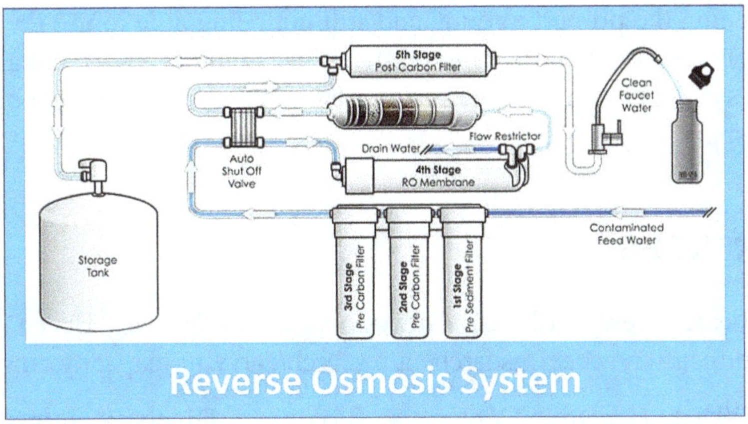

the .1 to 10-micron range, including bacteria. It does not remove dissolved contaminants that nanofiltration or reverse osmosis do. Microfiltration is performed by crossflows separation, which involves a feed stream being introduced into the membranes under pressure and passed over a membrane surface in a controlled flow path. High cross flow rates improve the ability of permeation. The separation of bacteria from water is common in many communities and municipal water treatment systems (WP, 2020).

The second first-hand observation and learning are tragic. The impact on the spring and groundwater is significant. Being that this is a historical site,

Figure 2. Microfiltration
(Water-online, 2022).

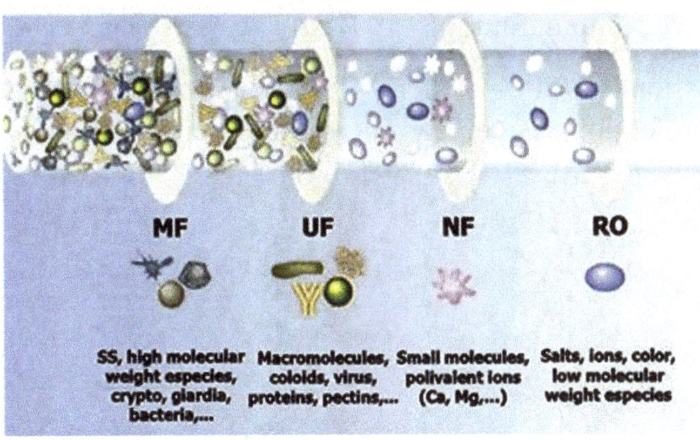

makes it even more heartbreaking. The number of elk on the property would have to be drastically altered to remediate the spring. This takes money away from keeping the property going, and still not being able to use the spring. I found out recently that they turned the property into a resort, put a reverse osmosis system in to purify the spring water, and now they have a beautiful inn and spa.

REFERENCES

Beef Research Center. (2020). *Carry capacity for grazing*. Beef Research Center. https://www.beefresearch.ca/research/carrying-capacity.cfm

CBD. (2020). *Grazing*. Center for Biological Diversity. https://www.biologicaldiversity.org/programs/public_lands/grazing/

FAO. (2019). *Livestock and Landscapes*. FAO. https://www.fao.org/3/ar591e/ar591e.pdf

Hilker, T. (2013). Overgrazing Turning Parts of Mongolian Steppe into Desert. Oregon State University.

Modderman, C. (2020). *Reduce Water Quality Issues from Manure*. University of Minnesota.

NWMC. (2020). *Water Resource Management*. National Water Management Center. https://www.nrcs.usda.gov/wps/portal/nrcs/main/national/nwmc/

Puretec. (2020). *Reverse Osmosis*. Puretec. https://puretecwater.com/reverse-osmosis/what-is-reverse-osmosis#understanding-reverse-osmosis

USDA. (2020). *Land Use*. Department of Agriculture. https://www.agriculture.gov.au/abares/aclump/definitions

Water Professionals. (2020). *Microfiltration*. Water Professionals. https://www.waterprofessionals.com/learning-center/microfiltration/

Chapter 7
When City Folk Wants a Tour

ABSTRACT

People raised in a city and going on to live their entire life in a city often do not fully understand nature, and the many nuisances that can happen. They may take tours, camping trips, or have guides show them the beauty of the great outdoors. Living and working with nature is different. It can be beautiful but also dangerous. Even the highly educated, who might think they are informed when it comes to nature, water, and sustainability, are often not informed. When with experienced professionals, it is important to listen to instruction; it may save your life. Through autoethnography, this chapter explores city folk and nature.

Water is the driving force in nature. – Leonardo da Vinici

MEMOIR

Class Action or Torte law was initially a way for people that have been injured or done harm (civil wrong) in which the courts could impose liability on the party doing the harm. This was meant for the greater good, but as with all things meant for good, the bad find ways to make easy money. More on this later in the book. For now, we are going to concentrate on a lawyer. We had a lawyer come visit us, he was hired to represent us with a class action law

DOI: 10.4018/978-1-7998-9691-3.ch007

suit. He was from New Jersey, a partner in a prestigious law firm, and grew up in New Jersey. Basically, he was a city boy. He had been on vacations, hike on trails, biked trails, but had never really been in what I call raw nature or untamed nature. Due to the law suit, he wanted to visit all spring sites, and experience them in their natural state. We told our corporate lawyer we would take care of him.

He arrived in dress slacks, dress shirt and expensive shoes. I took one look, and thought, "do I try to buy him some boots and jeans"? He said he had a plane to catch at the end of the day, and asked if we could get going, I explained to him the terrain we were going to would not be good for his clothing and especially his shoes. He was a kind person, full of excitement, and said let's carry one. Oh boy! What was I thinking! We spent most of the day driving through woods looking at springs and tanker loadout stations, and I explained the geology. Our last stop was the closest to the airport, and I really thought he would not want to go into the woods. I hoped I had worn him out. He wanted to experience the wetlands, bogs and swamp that led to the spring. I got a wooden pole to carry with me, instructed him to walk between me and another geologist. I stressed the importance of stay on the path, and following my footprints. I described the make-up of the bog, how dangerous they are if you go in one, they look like grass, but you will instantly sink. A bog it is like quick sand, and you will not be able to get out without help. It can be long and difficult process to remove a person from a bog.

We set off, I heard a whimper, and the geologist in the rear yelled out to me. I turned around and our lawyers head was sticking out of the bog with a look of stark fear on his face. I rushed over with my pole, and asked him if he could move his arms to grab the pole. He "whimpers" and said he could not even move. Tears begin to form in his eyes. I am sure he thought he was gone. I told my partner, that I had to go in. We always carried a rope, so we tied the rope around me, I took my pole and went in after him. After a while I got him lose and we were able to get both of us out of the bog. We were both muddy, him head to foot, his shoes were lost for eternity and we smell like rotten eggs. We got him back to the plant and hosed us both off, and we decided we needed to take him to Macy's to buy him new clothes before he was to get on the plane. His clothes were ruined, one of the quality control technicians had gym clothes in their locker and offered them up to us. After a good hot shower in the locker room, we dressed him in gym clothes and headed to Macy's, and put him on the plane. I was the one left to call corporate. Oh joy!

I called our corporate attorney to give him the bad news, and inform him we almost did away with the lawyer he hired. He was not amused! The quality manager laughed so hard he was crying. The plant manager tried to stoic, but broke down. The guy who gave the gym clothes were literally laying on the locker room floor laughing. I was worried about his recovery. I was afraid we might have ruined him for life. In the morning I called the lawyer to check on him, and I could tell he was still shaken. He said he would not be back on a spring tour. He could not understand why I could just take loom and fill in the bog and surrounding wetland. He really felt it was too dangerous. I explained that wetland and bogs are good for the spring and spring recharge area.

Many people do not understand that you cannot just eliminate nature to suit your purpose. Many think it is okay to fill in wetland, marshes and bogs. Not only does disrupt nature, it is against the law. The United States Corps of Engineers in 2018 significantly altered nationwide permits for dredging or filling wetlands. This went into effect in June of 2020. Property with as little as on tenth of an acre is regulated under federal and state law (U.S. CE, 2020).

WETLANDS

The U.S. Fish and Wildlife Service is the primary federal agency that provides information to the public on the nation's wetlands. A national database inventories, tracks wetlands and habitats associated with wetlands. The wetlands status and trends project are based on a monitoring program. This is extensive! This program provided information on recent or historical changes to the nation's wetlands, deep-water habitat, locations, and trends to best support needs.

In the early 1600's colonial settlements were draining wetlands. In the 1600's the U.S. had approximately two hundred twenty million acres of wetlands. By the 1980's only one hundred three million survived. Six states lost eighty five percent or more of their original wetlands. Many of our historical events, technological innovations and what society values at a given time, can have consequences and destructive effects. As society becomes better informed views of the natural world have changed. The preservation of wetlands is more fully understood, and how the ecological impacts fish, wildlife and the environment as a whole (FWS, 2020).

During the 1700's wetlands were regarded as swamps the just bred diseases, restricted travel, and impeded in food production. Most early pioneers viewed natural resources to be used without limit, and restrictions to be stopped. This caused widespread drainage of wetlands. This mentality still exists to some extent today. Thus, the lawyer saying just drain the wetland and bog.

Much of our early knowledge of wetlands comes from maps and surveyors' notes. These records are few and in many parts of the U.S. they were no surveyors. The north was clustered, and surveyors were more common. Many of their notes are difficult to read and understand because they are all hand-written. In the south, communities and land holders were scattered. It was not until 1785 that the Land Ordnance Act established the U.S. Public Land Survey. However, in 1849 congress passed the first of the Swamp Land Acts, which granted all swamp and overflow lands in Louisiana for reclamation. In 1850 they added twelve other states. This was the land-use policy for the coming century.

The westward expansion after the civil war and the building of the railroads ruined several of the great wetlands, swamps and forest. Almost all of Ohio's wetlands, including the great Black Swamp, were destroyed, which was rich in elm, ash and wildlife. Great areas of wetlands laid in the way of the railroads, and all were destroyed. In other parts of the U.S. urban and rural areas small wetlands were being drained for large projects. In Florida, there were many attempts to drain the Everglades. Severe flooding was the result, and flood control systems were built (FWS, 1990).

In the 1960's most of the devastation had happened. The Federal Government subsidized wetland losses due to public works projects. Since the 1970's there has been an increasing awareness that wetlands are valuable areas that provide environmental functions, prevent flooding, protect groundwater and surface water, animal and fish habitat. In 1987, federal efforts began to restore wetlands. Today, attempts are underway to restore parts of the Everglades (FWS, 1990).

In a recent USGS podcast, Science in the Swamp, Outstanding in the Field, the science of Louisiana wetlands is a significant concern. Louisiana has experienced the greatest coastal wetland loss of any other state. About one quarter of the state's coastal wetlands have been lost since the 1930's. The Mississippi River carries sediment downstream to the Gulf of Mexico, it feeds wetlands along the way. Alterations to the river, sea level rise, and subsidence, are resulting in the disappearance of Louisiana's wetlands. USGS scientist recently estimated that Louisiana has lost one football field worth of wetlands every one hundred minutes (USGS, 2021).

Wetland are found at the intersections of land and water. They are the places that are always wet and always dry. There are many types of wetlands. They have water above or below the ground for part or the entire year. Special soils are formed under the wet conditions, and planet have adapted so that they grow in water or wet soil. Moreover, a place does not need standing water year-round to be a wetland. Groundwater can feed wetlands and produce stable water levels than those that are fed by surface water. The plants present will vary on water conditions. Deep water wetlands grow different plant communities than shallow wetlands. Furthermore, wetlands that become dry each year grow different plants that those with persistent water. Three primary factors produce different wetlands. Hydrology, soil type and plant life the site supports. The hydrology is based on timing, frequency and the amount of water on the site. Types of wetlands can be broken down into marshes, sedge meadows, forested wetlands, shrub thickets, bogs and fens and rare wetlands (FWS, 2020).

The USGS refers to sequestration or capture and storage of an element, like carbon, in the environment. Wetlands have a significant role in the global carbon cycle. Wetlands are highly efficient in accumulating soil organic matter and can function as important carbon sinks. As the wetlands disappear, so does the carbon storage for the planet. Scientist at the USGS Wetland and Aquatic Research Center study wetland ecosystems to understand their long-term capacity for carbon sequestration, how the ecosystems are responding to natural and man-made induced changes, and how these changes will affect coastal wetland's ability to sequester carbon in the future. In addition, scientist are studying how an abundant wetland plant Cattail's have increased in the U.S. due to human activity. Cattail's often change wetlands undesirably (USGS, 2020).

MARSHES

Marshes are deep water wetlands that have a depth over six inches to six feet. Most have standing water year-round, some for part of the year. Life is supported both from groundwater and surface water. Shallow marshes seasonally flood by capturing rain and snow melt, but will often be dry by the end of summer. Deep marshes may have high water levels most of the year, and are often next to lakes, rivers and springs.

Marshes recharge groundwater and streams. The presence of marshes in a watershed reduces the damage by floods by slowing and storing flood water. This is an important function for droughts. When water moves slowly through a marsh, sediment and other pollutants settle to the floor or the marsh.

Marsh vegetation and microorganisms use the excess nutrients for growth that can otherwise pollute surface water such as nitrogen and phosphorous from fertilizer.

Since non-tidal marshes are prevalent in the U.S., they are mostly freshwater, although they often look brackish or alkaline. They are frequent along streams in poorly drained depressions or shallow water pools. The water level varies seasonally. Highly organic rich soil and clay is at the bottom, with lily pads and cattails, reeds and bulrushes which provide habitat for birds and small mammals.

Marshes, mangrove forest, and seagrass beds exist in an equilibrium of sediment, erosion and deposits. Long term changes can affect the development and stability of the vegetated beds, which make the marshes a rich ecosystem. When the tidal range is large and can transport sediment quicky, then the sediment will be adequate for marsh elevation to keep up with sea level rise. Marshes will move upslope, then, upland is replaced by marsh and low marsh by open water (USFW, 2020).

SALT MARSHES

Salt marshes are coastal wetlands that are flooded and drained by salt water brought in by tides. The soil is peat and deep mud. Salt marshes occur globally, but are more prolific in middle to high latitudes. They thrive when shorelines are protected with common habitat in estuaries. Approximately half of the nation's salt marshes line the Gulf Coast. Their habitats provide food for healthy fisheries, birds, shrimp, crabs, and other wildlife. Salt water marshes protect the coast form erosion and hurricanes by buffering wave action and trapping sediments (NOAA, 2021).

Sedge Meadows

These meadows are filled with grass like plants called sedges, with wildflowers growing in them. They will contain the most water after spring rains and snow melt. By the end of summer little standing water will remain. For this reason, they are often call dry marshes. Sedge meadows are found between uplands and lakes, rivers or streams. Many have been drained for farming, and many have been destroyed by pollution, sedimentation or heavy grazing.

Figure 1. Salt marsh Maine
(Geologic Survey, 2001).

It will take hundreds of years without farming or settlement to get these back. In the meantime, the ecosystems have been destroyed.

Forested Wetlands/Swamps

Forested wetlands are often referred to as swamps. They are dominated by trees. They are often found next to river or prolific springs. They are common in northern forest, and are characterized by the presence of trees. The soils are typically wet in spring and early summer, but dry up later in the year. In some forested wetlands, trees will water marks that can be several feet up their trunks. Most of these trees will have shallow root systems.

Shrub Thickets

These are wetlands dominated by shrubs and small trees less than twenty feet tall. Periodic flooding, logging or wildlife keep shrub thickets from becoming forest. Typically, the soil in shrub thicket is saturated with water. They have an abundance of shrubs growing among grasses, sedges, and wildflowers. They grow along stream banks or between forested and open wetlands.

Figure 2. Bog
(Kennebec Land Trust, 2000).

BOGS

Bogs are uncommon wetlands that are formed from lakes, and collect rainwater and accumulate dead plant material to form a dense, spongey mat on top of the water with grass and moss. Often there are trees growing in it. They have peat because of the dead and decaying vegetation. Bogs support plants and animals who are adopted to low nutrients. They are home to a variety of insects and amphibians. A bog is formed when poorly drained, depressed areas fill with plant matter. These form layers that create peat. They also trap precipitation, that make a bog a freshwater wetland. Bogs are generally found in cool northern climates. Many bogs were formed by poorly drained lakes created when the glaciers melted. The world's largest wetland is a series of bogs in Siberia. The Western Siberian lowlands cover more than 386, 102 square miles (NG, 2020).

There are many kinds of bogs. Blanket bogs develop in highland areas which have high levels of precipitation, thus the name blanket, the entire area is saturated. Cataract bogs are ecosystems connected to freshwater streams. Quaking bogs develop over a lake or pond, with bog mats. The mats are thick layers of vegetation. These often have a top layer of vegetation that measures three to four feet thick. These bogs get their name because they are known to bounce when people or animals walk on them. Raised bogs are domed-shaped from layers of decaying vegetation that accumulates in the center. String bogs have a varied landscape, with low-lying little islands that seems to interrupt a saturated bog ecosystem. Valley bogs develop in shallow valleys where rainwater and snow melt accumulate. Bogs take hundreds or thousands of years to develop. As plants decay slowly in bogs soils are oxygen and nutrient poor which makes the soils more acidic. Fungi and low-lying shrubs, such as heather grows in what is called a "heath", which forms after the abundance of heather blankets them. These spongy layer form "Peatlands" (UFWS, 2020).

PEATLANDS

Peat is considered a fossil fuel, and is the first stage of plant material turning into coal. Ancient bog plants, mostly sphagnum moss, are the fossils in peat. People have harvested peat for thousands of years as an energy source. Thick block of peat is cut and dried, then the blocks are burned. This is still a fuel source for Western Europe. In Asia tropical bogs are vital sources for lumber. Often dried peat is use to increase soil's ability to retain water necessary for agriculture (NG, 2020). Bog ecosystems are home to many species of wildlife and crops. Cranberries, blueberries and huckleberries are grown in bogs. Insects thrive in muddy bogs and consume plants, fungi and pollen. Many of these species do not live any other ecosystem. Other wildlife is supported by plants and insects. Amphibians, birds and mammals make their home in bogs (NG, 2020).

Peat bogs are known to be carbon sinks, they store enormous amounts of carbon, a process call "carbon sequestration". This is a process where carbon dioxide is captured and removed from the atmosphere, then stored in a natural facility. Since plants are a major source of carbon in the environment, by creating new bogs, millions of tons of carbon are sequestered. It is estimated that two billion tons of carbon is stored in the world's bogs. Most of this is stored in semi-frozen tundra and boreal forest in the area around the north pole in Siberia and Scandinavia. As areas of tundra thaws, millions of tons of carbon are released into the atmosphere. Carbon is emitted as bogs are destroyed for development and peat extraction. Peat ignites easily and burns slowly. The fires in Peat will burn underground for months, releasing carbon into the atmosphere.

Draining a bog for development or peat extraction destroys the ecosystems. Unlike wetlands, bogs take thousands of years to develop and hundreds of years to recover. Just because they seem unsafe, it is irresponsible to draining and fill in a bog. Bog are delicate and fragile ecosystems. Many conservations efforts mostly focus on government protection and preservation of bogs. Bog restoration projects are rarer that regulations on development. Bog restoration involve two different approaches. One, the removal or partial removal of peat from the bog, while the second, maintains a crust of hard peat. Both require water saturation of the entire area, and introducing bog vegetation.

Ancient people viewed bogs as spiritual or haunted places. Often, they were the burial places for people. These bodies are often found and studied to determine the everyday life of ancient societies and cultures. Many did not die a natural death; some are found with nooses and arrows lodged in them. Some have been murdered or have been ritual sacrifices. The bog bodies are known for how well their bodies have been preserved. Similar to Egyptian mummies, some have skin, internal organs and beard stubble. Many have tattoos and unique hairstyles (NG,2020).

RARE WETLANDS

There are commonly two types of rare wetlands, ridge and swale and interdunal wetlands. Ridge and swale have dry sandy ridges alternating with wet areas. When the glaciers melted around thirteen thousand years ago, they would melt and re-freeze; then melt by shifting the weight of the glacier. This caused and uplifting, wave action, and varying lake levels over time this created dunes in shallow areas. Then alternating dry and wet conditions in the wetlands provide plant and animal diversity. Interdunal wetlands are low spots carved by high winds in the dunes. They are deep enough to capture groundwater. These wetlands provide critical habitat for many uncommon plant species and resting areas for migrating waterbirds (NOAA, 2020).

PROTECTION

The National Marine Sanctuary System (NMSS) serves as the trustee for a network of underwater parks encompassing more than 600,000 square miles of marine and Great Lakes waters from Washington state to the Florida Keys, and from Lake Huron to American Samoa. The network includes a system

Figure 3. Wetlands
(Maine.gov, 2000)

of 14 national marine sanctuaries and Papahanaumokuakea and Rose Atoll marine national monuments. They protect some of the most iconic underwater places throughout the U.S. (NMSS, 2021).

The Wetland Initiative (TWI) is a 501(c)(3) non-profit corporation that was incorporated in 1994. Their vision is a world with plentiful healthy wetland improving water quality, climate, biodiversity, and human well-being. The mission is to design, retore, and create wetlands. Through innovative, collaboration, and employ sound science to improve water quality, habitat for plants and wildlife, and our climate (TWI, 2021).

To TWI, wetland provide an array of valuable services for society and the environment. Not to mention they are beautiful places on earth. They combine two basic strategies. On the ground restoration of wetland landscapes both independently and in partnership with other organizations, and innovation of ways to promote and finance large-scale restoration by others (TWI, 2021).

National Center for Coastal Ocean Science (NCCOS) are studying the warner bottom water temperatures found along the continental shelf off the coast of North Carolina and how these temperatures are affecting the area's species composition. In addition, they are studying different habitat restoration techniques for salt marshes, seagrasses, oyster beds and coral reefs. Natural Centers for Coastal Science are using remote sensing and physical observation to provide coastal managers with the tools needed to protect public health, restore damaged habitats, and improve community interactions with other ecosystems (NOAA, 2021).

The Nature Conservancy (TNC) is a global environmental nonprofit working to create a world where people and nature can thrive. Founded in 1951, TNC has grown to become one of the most effective and wide-reaching environmental organizations in the world. TNC has too many projects to list, however, they work tirelessly of restoration and recovery of wetlands, and conservation of forest, protect all lands globally. They believe that "nature" is the best prescription for a healthy world. By grounding their work in science, they are supporting innovation, leadership, and communication for conservation success (TNC, 2021).

Education will lead the way to long term change. One organization has been a leader in preserving water resources, "Project Wet Foundation". Their mission is to advance water education to understand global challenges and inspire local solutions. They do this by publishing water education materials to different age groups. These materials work in any classroom, with any curriculum. By providing training workshops to educators at all levels, formal and non-formal, on diverse water topics. Educators will then have the ability to reach children with objective, experiential, science-based water education. In addition, Project WET have global water festivals to educate children in fun interactive environments by providing alternative learning opportunities including structured learning stations and exhibits where students engage in hands-on-water activities and investigations. These festivals are community driven and coincide on World Water Day or Earth Day (Project WET, 2021).

Globally, all living and non-living things interact with each other. This interaction forms an ecosystem. The ecosystem is interrelated and sometimes complex. So much so, that changing one element can impact the entire ecosystem – for good or bad. Because of this tapestry of living things, the long-term effects can be devastating. Thankfully, many state agencies and non-profits and concerned businesses are collaborating to save our natural resources. Through education and grounded science, the public can be informed and persuaded to change. I have not met a single person, when educated on the science, does not want to do the right thing.

REFERENCES

NG. (2020). Bogs. *National Geographic*. https://www.nationalgeographic.org/encyclopedia/bog/

NOAA. (2021). *Salt Water Marshes*. NOAA. https://oceanservice.noaa.gov/facts/saltmarsh.html

Project WET. (2021). *Vision and Mission*. Project WET. https://www.projectwet.org/about-us

State of Wisconsin. (2020). *Wetlands*. State of Wisconsin. https://www.wisconsinwetlands.org/learn/about-wetlands/wetlands-101/

The Wetland Initiative. (2021). *About*. The Wetland Initiative. http://www.wetlands-initiative.org/mission-and-vision

U.S. Army Corp of Engineers. (2020). *Permitting*. U.S. Army Corp of Engineers. https://www.usace.army.mil/missions/civil-works/Regulatory-Program-and-permits/Obtain-a-Permit/

U.S. Fish and Wildlife Service. (2021). *Wetlands*. U.S. Fish and Wildlife. https://www.fws.gov/wetlands/data/mapper.html

USGS. (2020). *Wetland Word: Sequestration*. USGS. https://www.usgs.gov/news/wetland-word-sequestration

Chapter 8
Valentine's Day Massacre

ABSTRACT

Sometimes the biggest voices are the ones who are the most uninformed and the easiest to manipulate. Those who are doing the manipulation have an agenda of their own, often not considering the greater good of others. Business fuels our economy, and many have made commitments to do no harm. Often businesses are attacked just because the wealthy see them as an inconvenience. Often, we are told that business is bad, corporations are harmful, and some are. Among the larger businesses, there are a few bad actors. Most are made up of employees just trying to grow the business segments they are in charge of. They work hard, care about their work, and want their efforts to positively impact the communities they are work and live in. As an employee, the author takes great pride in her work and desperately wants her work to be considered noble and worthy. She has made a good income, but she is a middle-class working person, just like most of people she works with. This chapter explores the author's experience with the Valentine's Day Massacre.

Water belongs to us all, Nature did not make the sun one person's property, nor air, nor water, cool and clear. – Michael Simpson

DOI: 10.4018/978-1-7998-9691-3.ch008

MEMOIR

The water business was growing at a fast pace. If we could not meet the demand our customers would go elsewhere. We had a timeline to build a plant in the Midwest due to the growth in this area. I was excited because the areas we were looking at had been hit hard with job losses due to manufacturing going overseas. We had two of our finest geologists looking for water sources, and they found the perfect site. The quality of the water and recharge areas were pristine. Most importantly the area had a significant river that had not been monitored or modeled. We brought this technology to all our sites. Using remote sensing and modeling we would know everything about the area above and below ground. More on this later.

We met with town officials and they toured some of our other sites. We met with the state Geologic Survey and the Department of Environmental Protection agency. They in turn checked our reputation with other state agencies and were informed we were the best. Reputation is everything. This is why it is important to do things right, and if you mess up, admit it and fix it. This reputation is why we won awards from state and local non-profit agencies for our work.

When building a new facility, we would create an informational website. Then we would meet with the zoning and planning departments. Finally, we would have informational meetings with the public, which we would set up in a grade school gym and sometimes classrooms. They would fly several of us in to help. Unfortunately, it happened to fall on Valentine's Day. I had never been away from my children and husband on this day. I left two days before with notes for my husband; school valentines were in bags with names on them, and children's valentine heart-shaped boxes were in our bedroom ready to give out. A card for my husband promising a nice dinner when I returned.

The state geologist from my home state flew with me to help. A national geologist from the U.S. Geologic Survey volunteered to come and go over this state's geology. This was very generous; scientists do love to teach. Geologists in general are like this, it is in the fiber of our being. Geologists are all very laid-back people who love life -long learning. This national geologist felt that this state's earth science programs were behind the rest of the country's so he was excited to come. He knew of our work and was glad we were looking to establish a plant as this state desperately needed jobs and quality businesses.

Our Vice-President was coming, as was one of our plant and production managers. They would be in a classroom going over the plans, jobs, benefits, and pay. This new facility would add two hundred fifty high-paying jobs to this community. The two geologists that did the exploration work were in another classroom going over the site and the long-term monitoring program. The U.S. Survey geologist had a table in the gym, along with me and the state geologist from my state. I would answer questions, give out information and talk about permitting and compliance. We were proud of the work we were doing and felt it was our best informational meeting we had put together.

We knew the area was a favorite vacation spot for hunters and urban dwellers from Chicago. They had small hunting camps, cabins, and vacation homes on the area lakes and the river. According to a town supervisor they were not involved in the community. From the start of our project, our two geologists on the ground were visiting area residents in person to answer any questions and get a feel for the community. All were favorable, except one woman whose husband had an auto garage next to the river. She was a loud voice! One of the town supervisors informed us that she was mad at the town and state due to being fined for dumping motor oil in the river. To her, they owned the property and could do as they wanted. Unfortunately, you cannot dump motor oil into a river. It is a federal and state violation. It kills fish, wildlife, and plant life in a fragile environment.

We set up and were excited to show the community who we were. Little did we know that an ad ran in the paper of a large urban area an hour away advertising for paid activists. These are people who do not live in the community, have no vested interest in the community, but are paid to come and disrupt an event or meeting. Their pay scale determines how well they disrupt. They came, and they disrupted. They were paid by two wealthy Chicago attorneys, who also came and brought with them a Chicago television reporter. I ended up feeling so sorry for this young female reporter, she was totally out of her element. At this point in my career, I was being moved into corporate affairs and had undergone intensive media and crisis management training. I had already experienced a situation that I will go into in another chapter, so I was prepared when they stampeded to the gym, where they were directed first.

The two Chicago attorneys and reporter along with an elderly lady headed for me first. They come up to my table. The reporter had her microphone out and pointed at me. The little elderly lady shook her finger at me and said, "you're going to dry up all the wetlands". I hated to do it, but she really left me no choice. So, I asked her, "Ma'am, what is a wetland, can you describe it to me"? She melted. She had no idea; she was just saying what she was instructed to say. A guy who walked up with them, said very loudly, "I'm with the local Rivers Keeper group and I am concerned". By this time the attorneys were grinning big. I asked him what kind of metering he had done on the river and if was he doing any modeling. He said he had none and they were not doing modeling. I took the reporter's hand, brought the microphone over to me, and as loud as I could say, "you mean to tell me you represent a group trying to protect the river and you are not even monitoring it, what are you doing with all of your donations"? At this point, the two attorneys stopped grinning and took off to harass elsewhere. The riverkeepers guy came back around the table to talk to me.

I asked him why he was doing this. I was already planning on putting staff gauges and monitoring wells next to the river. I would actually have data that would protect the river long-term. He was embarrassed and told me one of the attorneys was his father-in-law, and if he wanted to stay happily married, he had to play along. I let him know their actions were putting the project at risk and hoped his rich father-in-law would donate two hundred and fifty thousand dollars to monitor the river. He took off and I never saw him again.

I glanced over to the U.S. Survey Geologist and he looked faint, so I sent our state geologist over to help him out. Our state geologist was a seasoned well qualified and experienced professional. He saved the day! There was a group of young arrogant and loud men that were just harassing this U.S. Survey Geologist saying they would bomb his building for coming to this meeting. The older more experienced geologist said pictures were being taken of them and they would be forwarded to the FBI. Then he asked them how much they were being paid, and that is when we learned that starting pay was $15 an hour, including the bus ride, but if they disrupted and caused confusion it would be $20 an hour.

Back where the Vice President, plant manager, and production manager were holding their meeting they tried the same tactic, but when they learned about our pay scale, they stopped harassing and started asking questions. Production line employees started at $17 per hour with benefits, good benefits. Blow molders made $21 an hour. Even the janitor made $20 an hour. Maintenance mechanics made $30 an hour. This was in the '90s so this was better than average pay at the time. We offered college reimbursement and offered adult classes in our lunchrooms. Our plants were so clean, you could eat off our filling room floors. The environment was nice and we had a low turnover in our facilities. People liked working for us.

Back where my buddies were, it was a massacre. They had a lady standing on a chair yelling at them. People were throwing things at them, to the point that they could not even talk. There was so much yelling no one could hear them. I mentioned earlier that geologists are a gentle kind lot, who love educating people and talking about the earth. The happiest group of geologists is when they are on a field trip with a cooler full of iced beer. This was not their domain. My heart broke for them. At the end of the night, they were battle weary.

In the end, it all came down to two rich men who did not want to see a plant or trucks in the community when they came into town to spend a week at their hunting lodge. They got a son-in-law, an out-of-compliance business owner, the one who puts motor oil in the river, to help them find out information. Then they put an ad in the paper to recruit people who thought they were earning easy money and rented two buses. All to destroy a meeting. Then they went back to Chicago and started filing lawsuits. By the way, the television station never broadcasted the story. Sadly, our company was not in the mood to keep fighting lawsuits when the plant could be built at a second site. So, we built on the second site in a neighboring state. I felt terrible for the town and supervisors.

COMMUNITY IMPACT

Again, I am going to say that not all corporations are bad. Some of the biggest have the funding to do great work. If we would have gone into this town, we would have been very involved in the local schools and non-profits. The state agencies knew this because they did a great job researching who we were. I have built science rooms, climbing walls, and ropes courses, and given money for all sorts of educational purposes. We worked with Junior Achievement in each of our locations. In one location we had groves of maple trees. We had a student organization from the local high school that each year tapped the trees, and made maple syrup. They designed their own labels. They sold their maple syrup at the school fair and mall. Each year they learned how to run a business, market that business, and how liquidate inventory. Most of these kids were not athletes, but kids who needed a purpose.

These plants are taxed. They hire local contractors, electricians, plumbers, and food service representatives. The average new facility was estimated to impact the local economy to the tune of twenty-four million dollars in the first year of operation. In five years, it was estimated they would impact the local economy by fifty-three million. Business is what drives the economy, it is what keeps our quality of life. Without thriving businesses, areas grow stagnant, lack jobs and people's quality of life disappears into poverty. No amount of free government money can make it right. The lawyers ended up being the most hated land owners in the community.

THE RIVER

It is almost unbelievable to represent a non-profit that is supposed to protect the rivers in a state and refuses to help install equipment that would model and protect the river for hundreds of years. I truly felt sorry for this young man, because his spirit to do good had been stripped from him because of a rich and powerful family member. He must have been terrified of his father-in-law. Eckart Tolle says that "the means to the end is so important to keep from corrupting the end." This means that if your "means to the end" you want is based on lies and unethical behavior, your end will not serve the greater good. This mighty river was worth protecting, I only wish we could have been a part of its daily life.

Dumping motor oil on a daily basis into a river can have catastrophic effects. When commercial garages, homeowners, or individual cars leak oil, they contribute to the stormwater runoff, then precipitation on parking lots and roads runs off into local water bodies. When oil is dumped into a river directly, it is in high concentrations. Oil is a pollutant and contains toxic chemicals such as lead, benzene, zinc, magnesium, and nitrates. In addition, when plastic debris, cigarette butts, paints, solvents, sediments, and other hazardous waste are dumped in the environment they are pollutants. Topsoil and natural vegetation can filter the pollutants out, but the roads and pavements cannot. So, runoff from roads and parking lots goes into the storm drains and directly into waterways. Dumping oil into a river is poison to all marine life and wildlife. It affects the entire ecosystem.

This does not just affect surface water, these pollutants seep into the groundwater supplies and contaminate the drinking water wells for homes and businesses. Upward forty percent of oil pollution in the U.S. comes from improper disposal of used motor oil by individuals. The U.S. Environmental Protection Agency (EPA) has specific advice for automotive maintenance to prevent and repair leaks and disposing of used motor oil and other automotive fluids. Batteries are included in this advice. Designated drops off or recycling stations are in every state in several areas. Commercial car washes are mandated to have catchments to handle oil and fluids. It is also

Figure 1. Water festival
(Project Wet, 2000)

suggested to eliminate or cut back fertilizers and pesticides for home use, such as on lawns and gardens.

By over-watering your lawn, you can create extra run-off as well. And if you are embarking on a residential landscaping project, try to incorporate permeable pavement (which allows run-off through it into the soils below) as well as rain barrels to collect water, and rain gardens, grassy swales, and driveway-side vegetative strips—all planted with region-appropriate native plants of course—to help filter contaminants out before they hit the storm sewers.

WATER FESTIVALS

One of the proudest projects I was fortunate to start and be a part of was nationwide water festivals. I worked with a great organization called Project WET (water education for teachers). This organization had coordinators in every state, who worked with the state agencies and teachers to educate children and the public on the science of water and sustainability. We started doing water festivals throughout the nation. They are still going on today,

Figure 2. Groundwater model
(AZ, gov, 2000).

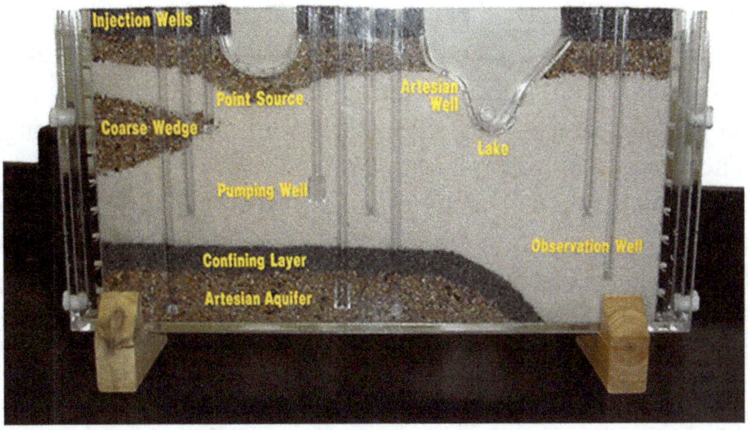

and thousands of children and adults have attended where they learned the science of water and sustainability.

One of the most popular displays is the groundwater model. It was made from plexiglass, with a 1-to-2-inch space between, and was around two feet high and three feet wide. It stood on wooden blocks to keep it upright. In between the plexiglass layers was stratified geologic material representing the ground below the surface of the earth. Placed within this material was a deep well, shallow well, a lake, a pond, and a wetland. I had food coloring representing different contaminates. Red was chemicals, green fertilizers, blue pesticides, and yellow garbage. We attached a hand pump to the model. To run the model, you filled it with water, and the hand pump acted just like well pumps, pumping the water through the watershed. As you added the food coloring, and pumped like you would in a household well, the food coloring representing the contaminant flowed through the model (aquifer).

When teaching adults, we went back to the basics. So many of them are just not educated in water science. We would start with the water cycle, and a game of different blocks, that we would give individuals. These blocks would represent clouds, animals, oceans, rivers, plants lakes, streams, etc. We would have posters which represented each stop of the water cycle. A person would stand by the ocean poster, another by an animal. As they rolled the dice, they would travel around the posters. This may be very simple to an academic, but it is surprising how well this teaches the basics of the water cycle. This game was developed for grades 4-6th, but it was also a wonderful tool for adults.

After this we would have them pump in contaminants in our groundwater model. Last, we would educate them on short and long-term monitoring. I would have a modeling program running on my computer where they could experience first-hand, real-time input from monitoring points throughout an aquifer. There were staff gauges in streams and rivers monitoring flow levels, Piezometers which are 2 to 4-inch wells at different depths that monitor levels in groundwater, and precipitation gauges. We would follow this with copies of water testing to show how water quality is important for the long-term sustainability of the watershed. We wanted to show how all life-forms are dependent on the watershed.

The rich lawyers were only concerned about the ability to come to a quiet area, and not see any activity. They were being selfish, and the community lost out on a great partnership. All partnerships have to be based on trust and openness. Education leads to good land use practices. We have just begun to understand the inequities of water access and more is needed to research and solve these issues. According to Patel and Lee (2017) is has to start with trust and tap water utilities have to include education. According to Patel and Lee's research, in the U.S., nearly one in four children do not drink tap water. Up to fifty percent of U.S. schools do not provide free water in school cafeterias (Patel and Lee, 2017).

According to Yolanda McDonald and Nicole Jones (2018) environmental injustices are a function of environmental discrimination. This can arise from industry and non-industry practices, policies and regulations. In many cases this involves groups that are under-represented. An example of this is in Flint, Michigan, where an under-represented economically depressed population was affected tragically by government officials not responding to water quality data and health concerns.

Going forward more research is needed by scientist and scholars alike over water management, long-term monitoring, remote sensing, and the modeling of groundwater and surface water. This research is mostly done by government agencies. Most scientists focus on a specific interest. McDonald and Jones (2018) had the same concerns noting three key initiatives needed to advance water quality for all groups. First, is education and awareness. Second, civic engagement. Third, public health practitioners need to be involved and use water results as a public health concern (McDonald & Jones, 2018).

REFERENCES

Eckehler, J. (2008). How to Keep Waste Oil out of a Waterway. *Scientific American*. https://www.scientificamerican.com/article/how-to-keep-waste-oil-out/

EPA. (2020). *Water Topics*. Environmental Protection Agency. https://www.epa.gov/environmental-topics/water-topics

McDonald, Y. J., & Jones, N. E. (2018). Drinking Water Violations and Environmental Justice in the United States, 2011-2015. *American Journal of Public Health*, *108*(10), 1401–1407. doi:10.2105/AJPH.2018.304621 PMID:30138072

Patel, A. I., & Schmidt, L. A. (2017). Water Access in the United States: Health Disparities Abound and Solutions Are Urgently Needed. *American Journal of Public Health*, *107*(9), 1354–1356. doi:10.2105/AJPH.2017.303972 PMID:28787195

ProjectWET. (2021). *Programs*. Project WET. https://www.projectwet.org/

Chapter 9
The Bear and Eagle

ABSTRACT

Water is the most valuable natural resource on the planet, and after that, it is the old growth forests. For this, our partnerships with other parts of the world should be a major focus. The United States and Canada have a long and unique relationship. This relationship is based on shared values, geography, common interests, deep personal connections, and multi-layered economic ties. Together, these economic ties have created the largest trading relationship in the world. A secure and efficient flow of goods and people across the border is paramount to both countries' economic prosperity. By working together, they enhance their security and accelerate the legitimate flow of people, goods, and services. Both countries remain committed to close cooperation on issues facing both countries and global challenges. Through autoethnography, this chapter explores this relationship.

Gene Roddenberry continually reminded us that the Star Trek Enterprise was a metaphor for starship Earth. And the strength in this starship came from its diversity, coming together, and working in concert as a team. That is the strength of our countries, Canada, and the United States. We are nations of diversity. – George Takei

DOI: 10.4018/978-1-7998-9691-3.ch009

MEMOIR

Canada was part of my territory. Which I loved, because I love all things Canadian, including my husband who is French Canadian. I was spending an enormous amount of time in Canada searching for springs, helping with permitting and the renovation of a plant. I had a technician who was helping me in the Montreal area, and he was wonderful. A very nice man was good at his job, and I knew I could trust him. He became a good friend to my husband and me, and I enjoyed many conversations with him. One in particular!

He told me he believes that a country's emblems were inspired by God, and truly represented the culture of the country. Such as the Bald Eagle is the emblem of the United States. In 1782 the Eagle was chosen to represent the U.S. because of its long life, great strength, majestic looks, and freedom. The eagle is believed to be the chief over all winged creatures. The strength represents authority with strong wings, strong legs, and talons. Spiritually, the eagle has been seen as representing the all-seeing eye. In addition, it represents release from bondage, victory, speed, and royalty. Because of this, it is often a symbol of powerful nations. In 2016, the Bison or Buffalo was named as a U.S. symbol. The bison is a symbol of resilience, shared values, a healthy landscape, and communities.

When researching for this book, and according to the Canadian heritage office, the beaver and the Canadian horse are the national animal symbols of Canada. According to Canadian websites, the beaver has been the symbol for Canada, since 1975 and the Canadian horse since 2002. However, a Canadian senator has launched a campaign to replace the beaver with the polar bear.

As you learned in chapter one, I love beavers. They are kind and live to play practical jokes on rehabilitators, and they are known as industrious master builders. The master building attributed to them, includes wetlands, bogs, and swamps, which in turn clean and preserve watershed areas. The indigenous peoples believed beavers to be the "sacred center" of the land. The Canadian Horse is a strong well-muscled breed of horse, usually dark in color. They descended from draft horses that were imported to Canada in the 1600s. The polar bear is noted for its strength, courage, resourcefulness, and dignity. It is often called Canada's most majestic and splendid animal. In addition, it continues to be a powerful symbol in the indigenous people's culture. People often say it reflects the soul of Canada.

Canadians and Americans share a common belief; our independence and freedom are the core of our values and culture. We both share strength and love the majestic land that is our birthplace and home. Part of that is honoring our water, trees, and environment. It does not mean that we have not had problems in our past, and problems today. That strength, values and beliefs will guide us through and lead us to a better quality of life.

One such issue for Canada is the Province of Quebec periodically wants to secede from the rest of Canada, just as the U.S. southern states did at one time. There is a divide of ideology in both Canada and the U.S. One thing we both hold dear is the protection of our natural resources. Hopefully, the bond between the U.S. and Canada is never broken, for we must continue to be friends to support liberty and freedom globally. I brought up the secession of Quebec because it is an important part of my memoir.

I realized that traveling to Toronto or Ottawa was very pleasant and people were always friendly. However, traveling to Quebec and Montreal was starting to become difficult, and sometimes I was worried about my safety. I would hear whispers in restaurants about the American over there eating, we do not want her here or go home American. My husband said he would go with me to Quebec Province whenever I went because his husband being French Canadian, felt it would make a difference. What a difference it made, speaking their language, they all loved him. Returning to the same hotel and restaurant, I was treated like a queen.

My husband filled in as a second field technician. One day we went to take water samples from a tanker truck fill station and spring house in the middle of nowhere. It was a small concrete house (with six-inch pipe), with a window at the very top, both of us could barely fit inside. The port to fill the testing bottles was in the six-inch pipe. I had my husband hold the bottles, and I would fill them. As I turned the valve to start filling the bottles it broke off, and suddenly water started gushing out at two hundred fifty gallons a minute, filling the concrete house like it was a water tank, and the pressure was so strong we could not open the door because the water was hitting the door. I must admit that I started to panic thinking that our children were going to lose both of their parents that day. The electrical panel box was at the very top, it had to be turned off. With water filling up the structure, I really thought we would be electrocuted turning off the pump. I stood on a pipe and climbed up to turn if off as my husband held me and said a silent prayer. Fortunately, I was able to shut the electricity off and stop the pump without either of us hurt.

On that day, my husband got a new appreciation of my work. Those that manage natural resources and scientists in the field are subject to all kinds of weather conditions, animal attacks, ticks, bee stings, snakes, and equipment failures. In addition, one must be careful of quicksand, bogs, and swamps. Protecting our great places is hard work, but worth it.

After completing our sampling, we met with the people renovating the plant. We decided to ship water via tanker truck to Maine, to be bottled and shipped back to Canada. The spring was a mineral spring and very high in Total Dissolved Solids (TDS), and I was excited to run it through our microfiltration system. We received our first loads and were surprised at the horrible aroma of the water. The quality control technicians were aghast because of the sulfates. We ran it through our filtration system, bottled it, and shipped it back. The filtration made such a difference, we were so proud to ship back such a great product.

In a few days, we received a call asking us why we ruined the water. The water was popular with Canadians because it was considered mineral water and a laxative. They went elsewhere to be bottled. In the U.S. we think we do it the right way, but not all people look at basic things as we do. Even those neighbors closest to us. It does not mean one is right or wrong. It is what it is, and we must be accepting of each other. In the U.S. we have specific tastes, especially for our water. We want it pure, clean with low dissolved solids. It fits our palette better.

WATER CLASSIFICATIONS

There are different classifications of water. The FDA and state departments that regulate and permit food items, see bottled water as a food product, and it is permitted under the FDA. The EPA classifies water as drinking water and classifies it as a utility. States have specific departments that regulate utilities and water resources. Water is one of the most regulated substances in the world, mostly in the U.S. and Canada. The World Health Organization (WHO) classifies water much differently than the U.S. and Canada. The U.S. is regulated by the Safe Drinking Water Act, which sets the standards for water systems. There are primary and secondary water standards. These standards are over microorganisms, disinfectants, by-products, inorganic chemicals, organic chemicals, and radionuclides (EPA, 2021). My yearly water testing included all of these and cost two hundred and fifty thousand dollars. Daily and monthly samples were taken for Total Dissolved Solids

(TDA), PH levels, microorganisms, and organic and inorganic compounds. The global stage is much different.

According to a study by Dunn, Bakker, and Harris (2014), Canada has similar issues to the U.S. They have great standards, where individual water utilities hold the responsibility to test and maintain water quality, but they have water inequity issues in areas of high poverty. The baseline guidelines are what is needed but the decentralization may create inequity. Throwing money at a problem and hoping you fix it, maybe counter-productive due to local politics. To really fix the problem, is to better educate society.

The WHO has no standards for TDS (total dissolved solids), chloride, sodium, potassium, hardness, and sulfate and their effects on health. The only standards established on

taste. Moreover, increased concentration of the above compounds will often lead to corrosion of water pipes or scale deposits. The nitrate and fluorine standard, however, is based on health effects. The chloride standard is based on a concentration of 250mg/l, with a concentration up to 600mg/l as safe. A high chloride concentration leads to the corrosion of metals in the distribution system, which is also related to the hardness of the water. Which may result in a high concentration of metals in drinking water. Sodium in high levels could cause high blood pressure, although no firm evidence supports this. WHO provided no data on TDA, but states that a TDS lower than 1,000mg/l is acceptable. Anything higher is considered corrosive. Calcium is somewhere between 100-300mg/l and magnesium is less than 500mg/l. As for Ph and Alkalinity, a hardness of more than 200mg/l will lead to scale deposits in the piping system. Soft water with hardness less than 100mg/l may cause corrosion in the system. Sulfate is one of the least toxic anions. High concentrations may lead to dehydration, stomach complaints, and diarrhea so higher than 500mg/l is not recommended. The WHO standard for fluorine of 1.5mg/l, anything higher may cause skeletal deformation (Van Der, 2003).

A five-nation member initiative between Germany, France, Netherlands, United Kingdom, and Denmark has agreed to collaborate to test products of drinking water. As part of a collaborative harmonization, nations such as Germany, France, the United Kingdom (along with Ireland), and the Netherlands announced in January 2011 that they have formalized arrangements to work together on the regulatory frameworks to ensure the safety of drinking water. Since the formalization of this work the membership of the 4MS-Initiative is as follows: (Federal Ministry of Health, 2021).

- The acceptance of the constituents used in materials in contact with drinking water.
- The testing of materials
- The use of common test methods and setting acceptance levels.
- The specification of tests to be applied to products.
- Reviewing factory production control and setting audit testing requirements
- Assessing the capabilities of certification and testing bodies

The purpose of this collaboration is that water is considered an important food product, used for drinking, cooking, and bathing. The purpose of this collaboration is to provide stringent quality standards for drinking water. Their Drinking Water Ordinances specify requirements in which drinking water must not only be free from unsafe levels of substances and pathogens but that its must be pure and wholesome (Federal Ministry of Health, 2021). In addition, this collaboration specified that the drinking water ordinance should regulate water suppliers and monitor authorities.

WATER CLASSIFICATIONS

Mineral water is considered water that comes from an aquifer that comes to the surface under its own pressure as a spring, or that comes from a drilled well to remove the water. It must be microbiologically safe, and the only treatment allowed prior to bottling is the removal of any iron and sulfides at high levels and to the addition of carbon dioxide for carbonatization. Mineral water is usually bottled at the source, more so in Europe than in the U.S. and Canada. Mineral water is considered to have specific health benefits. The Canadian water bottled in the memoir was mineral water (Van Der, 2003).

Of course, there are varying parameters in different countries. There are two hundred ninety-one mineral waters from forty-one countries for sale on the retail market. Most of these originate in Europe. In addition, mineral waters that originate in Europe are sold in Asia, Africa, U.S., Canada, Australia, and Central and South America. Five countries sell more than ten different mineral waters and these are Germany, Italy, France, Spain, and Belgium (Van Der, 2003).

As with all bottled water, the analyses and points of bottling are on the labels. This is very important in mineral water because the contents can be very different according to the geology of where the water originated. Most

Figure 1. San Pellegrino
(Nestle, 2022).

mineral waters have a TDS of 50 to 500 mg/l. Generally, they fall closer to the 500 level. Mineral waters will contain, bicarbonates, sulfates, chloride, calcium, magnesium, fluoride, and sodium. Depending on the country, the acceptable levels may change. In the U.S. from 250 to 500 mg/l TDA is usually classified as mineral water (CDC, 2021).

Spring Water

Spring water comes from an aquifer and flows naturally to the surface under its own pressure. Unless it is removed by what is called a borehole which is a well that correlates with a spring, both in chemistry and geology. This is done to meet the packaged food requirements for HACCP (Hazard Analysis Critical Control Points). This program is standard among packaged food companies to safely process food. There are seven steps to the HACCP program: (FDA, 2021).

- Perform a hazard analysis.
- Determine Critical Control Points (CCP's)
- Set Critical Limits

Figure 2. Poland Spring
(Nestle, 2000).

- Establish a monitoring system.
- Established corrective actions.
- Established verification procedures.
- Establish record-keeping procedures.

All water used in food production or to be bottled falls under these standards. This includes spring water.

Artesian spring water is water that comes to the surface on its own pressure or comes from wells drilled into the aquifer to remove the water. This is usually water from deeper levels and may come from fissures in the bedrock. It usually has a higher TDA level and may be mineral water. When tapped, the pressure in the aquifer, commonly called artesian pressure, pushes the water above the level of the aquifer, sometimes to the surface.

Figure 3. Purified water
(Members Mark, 2022).

Purified Water

Purified water is water where everything has been removed from the water except for small amounts of minerals. This is usually done with microfiltration or reverse osmosis. This is generally considered water for pharmaceutical purposes.

Distilled Water

Distilled Water is water turned into a vapor, leaving minerals behind. The vapors are then condensed into water again. This is done with a distillation unit. These are often turned into a de-salination unit. Distilled water is used for iron, and other equipment that you do not scale to build. De-salination units are used to turn sea water into a drinkable water.

Figure 4. Distilled water
(Fischer Scientific, 2022).

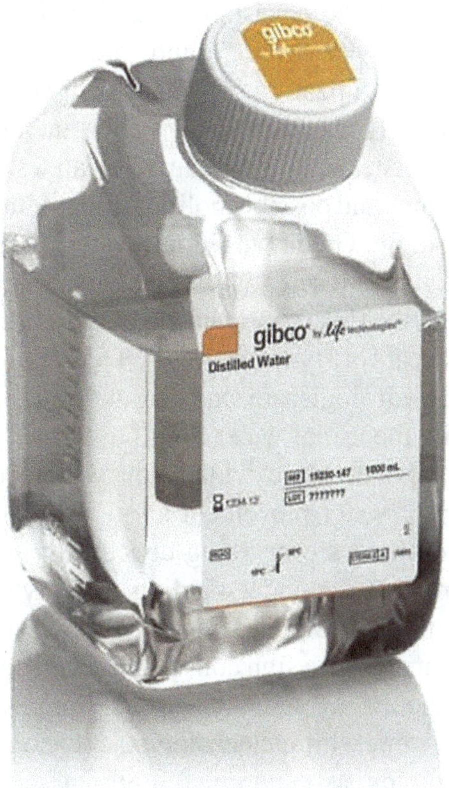

SCIENTIST WHO MANAGES WATER AND SUSTAINABILITY

The biggest water problems worldwide are quality and scarcity. We can go from one area that is lacking in water to another that is poor in quality. Some nations want to grow industry, but in the process leave certain groups marginalized. When industry up stream is taking so much water that the farmer downstream cannot irrigate crops or water livestock. Land use practices have impacted water quality and deforestation has reduced the groundwater levels in some areas significantly. As the BRIC (Brazil, Russia, India, and China) nations industrialize and grow there is a concern that water shortages will only grow, and water pollution will lead to massive devastation (Jury and Vaux, 2005).

Water shortages and loss food production will rise dramatically in the next two decades for many countries. It is already projected that India and China will be first and this could be devastating for their large populations. The presence of persistent and pervasive ground water overdraft, the specter of climate change, and the salinization of soils in arid and semiarid areas will exacerbate the problem. This brings about a continued growing demand of food imports, and upstream and downstream riparian water rights, which will bring economic hardships and possible political unrest (Jury and Vaux, 2005).

Hydro" comes from the Greek word for... water. Hydrology is the study of water and hydrologists are scientists who study water. The discipline of hydrology has evolved, the first report written by the USGS was back in 1986 titled "Two Hundred Years of Hydrogeology". Today is a growing discipline, of the study of water. Hydrologist are studying and finding solutions to water problems, maintaining the use of water for all users.

The hydrologic or water cycle which you will read about in chapter fifteen, is the system of water movement around the planet. Hydrology is the science that encompasses everything concerning water on our planet, from the transport processes to understanding quantity and quality of water as it moves through the hydrologic system. Water resource managers are involved in planning, analysis, design construction and operations of projects involving water and the management of water resources. With water issues growing, hydrologist work with meteorologist, oceanographers, geologists, chemists, physicists, biologists, economists, political scientists, specialists in applied mathematics and computer science, and engineers in several fields (USGS, 2021). A hydrologist may plan multimillion-dollar projects or give homeowner advice. Water issues have a wide range!

The scientific field associated with sustainability is enormous and described as an emerging field of research dealing with the interactions between natural and social systems, and with how those interactions affect the challenge of sustainability while meeting the needs of present and future generations while reducing poverty and conserving the planet's life support systems (Katz, 2011).

This emerging field of research has experienced extraordinary growth and identifies geographical regions with unified scientific application and study. This field has a different field structure than other sciences but is not successful without a collaborative network. Scholars in the field might argue that sustainability science publications should contain integrated research and not include to offer perspectives or advocate policies, because of the different representations tied to institutions and governments. Traditional science is represented by Japan, the United States, and Western Europe, however, the emerging BRIC along with South Africa, Kenya, and Nigeria economies are developing may differ from the traditional views (Katz, 2011).

Sustainability science is reflective of the multidisciplinary that it spans from natural science, technology, and social sciences. That is, it unifies many scientists globally. This unification has led to sustainability science as an emerging academic field of study, when many felt it difficult to conceptualize, define, and measure sustainable development, therefore making it subject to scientific inquiry. The U.S. National Academy of Sciences created a Board of Sustainable Development and after five years of effort connected science and technology to create the science. In addition, further meetings were held at the World Academies of Sciences and Technology, to identify core questions and methodologies, and launch sustainability science globally (Katz, 2011).

The search for core questions and research themes was an effort to consider scientific literature and address human environments and nature. Combining these to come up with core questions and thematic research. There are seven major questions that were developed for research: (Katz, 2011).

- What shapes the long-term trends and transitions that provide the major directions for this century?
- What determines the adaptability, vulnerability, and resilience of human-environment systems?
- How can theory and models be formulated that better account for the variation in human-environment interactions?
- What are the principal tradeoffs between human well-being and the natural environment?
- Can scientifically meaningful "limits" be defined that would provide effective warning for human–environment systems?
- How can society most effectively guide or manage human environment systems toward a sustainability transition?
- How can the "sustainability" of alternative pathways of environment and development be evaluated?

Sustainability science has proven its usefulness, with tens of thousands of scholars who have researched and published journals, articles, and books. Not only has it inspired scholars, thousands of authors, practitioners, and knowledge seekers. Teachers and students have been inspired to publish, participate in sustainable projects, and have gained a greater knowledge of how to protect our planet, and our local environment, and develop solutions for the greater good and future of humanity. It is so important to keep this science interdisciplinary to ensure equality to sustainability so that not only the hard science is considered but also the soft science is incorporated and taken seriously. We cannot ruin our economies but must build on them, with good science and thoughtful solution-filled practices.

REFERENCES

Canadian Heritage Office. (2020). *Canadian Symbols*. Canada Government. https://www.canada.ca/en/canadian-heritage/services/official-symbols-canada.html

CDC. (2021). *Drinking Water Standards and Regulations*. Center for Disease Control. https://www.cdc.gov/healthywater/drinking/public/regulations.html

Dunn, G., Bakker, K., & Harris, L. (2014, May). Canadian Drinking Water Standards. *International Journal of Environmental Research and Public Health, 11*(5), 4634–4651. doi:10.3390/ijerph110504634 PMID:24776725

EPA. (2021). *National Drinking Water Standards*. EPA. https://www.epa.gov/ground-water-and-drinking-water/national-primary-drinking-water-regulations

FDA. (2021). *Water standards and classifications*. Federal Drug Administration. https://www.fda.gov/consumers/consumer-updates/bottled-water-everywhere-keeping-it-safe#:~:text=FDA%20describes%20bottled%20water%20as,the%20limits%20set%20by%20FDA

Federal Ministry of Health. (2021). *Drinking Water Ordinance, Umwelt Bundesamt*. Federal Ministry of Health. https://www.umweltbundesamt.de/en/topics/water/drinking-water/distributing-drinking-water/approval-harmonization-4ms-initiative#basis-for-a-common-approach

Jury, W.A., & Vaux, H., Jr. (2005). The Role of Science to Solving the Worlds Water Problems. *NAS, 102*(44), 15715-15720. doi:10.1073/pnas.0506467102

Katz, R. W. (2011, December 6). What kind of science is sustainability science? *Proceedings of the National Academy of Sciences of the United States of America, 108*(49), 19449–19450. doi:10.1073/pnas.1116097108 PMID:22114189

USGS. (2021). *Hydrogeology.* United States Geologic Survey. https://www.usgs.gov/special-topic/water-science-school/science/what-hydrology?qt-science_center_objects=0#qt-science_center_objects

Van Der, A. M. (2003). Classification of mineral water types and comparison with drinking water standards. *Env Geol, 44*(5), 554–563. doi:10.100700254-003-0791-4

Chapter 10
The Scientists

ABSTRACT

Most scientist agree the climate is changing, but accurate science is needed, and more research is needed to know what is really going on. Politicians and the public need to listen to the scientists. This is necessary for preparation and ethical decision making. The scientist is exploring the truth and finding how to navigate this new world. It is interesting that the scientific community is not divided. It is our media and politicians that are divided. This shared sense of purpose among so many scientists, non-profits, and communities is going forward without the ones yelling and screaming and ready to take down economies. The key is the work "communities" working with the scientists to bring about a better world and work toward the greater good.

The age of climate decisions is here, and our actions will define the course of civilization and the health of the planet. – Dr. Paul Mayewski, Director of the Global Climate Institute, University of Maine

One of my projects was to restore an historical spring and bottling plant. In that process, I advertised for historical memorabilia and artifacts. This historical spring and bottling plant was one time connected to a grand resort. It was founded in 1795 and by 1900 was an international resort for the rich and famous, and the spring was known for its curative powers. I purchased several brochures and one of these brochures was dedicated to the ice storm

DOI: 10.4018/978-1-7998-9691-3.ch010

of the century which was in 1898. In 1998 we had another ice storm of the century. People occasioned this as a product of global warming and the earth was doomed. This was a hundred-year weather event. Global climate is changing, and it is important to understand the difference between a hundred-year weather event and global climate change.

Most scientists agree the climate is changing, but accurate science is needed, and more research is needed to know what is really going on. Politicians and the public need to listen to the scientists. This is so necessary for the preparation and ethical decision-making. Scientists are exploring the truth and are finding how to navigate this new world.

MEMOIR

I was fortunate to have several mentors that are in the acknowledgments of this book. I do not see myself as a scientist but as a project manager. I considered myself a good project manager. I was fortunate to hire great scientists who ensured our studies and reports were professional and scientific both nationally and internationally. It is largely due to these people that my tenure in my job, had absolutely no issues. We had a great reputation for doing great work and protecting our sources. This was because I listened to people who knew more than me. I learned and cared deeply about the watersheds we were managing.

The scientist that I worked with daily were working in consulting companies. I only used those who had great reputations and had been doing the work for several years. They were the best in their fields. In addition, I valued the partnerships I built with universities. When private and public institutions come together great ideas are shared, creativity is heightened, and educational and research projects can serve the greater good.

I worked with many research universities on projects relating to natural resource sites, and educational exhibits. It was a partnership benefiting all parties. The most interesting project was partnering with universities to build educational exhibits. These were a climate and weather exhibit, a glacial exhibit, a long-term monitoring exhibit, and a groundwater model and real-time weather station. In addition, we built cultural exhibits to show how climate and weather sustainability impacted communities. It comes down to individuals practicing leadership to make their community and state sustainable.

I restored a historical bottling plant and spring house. This building was on the historical register, so everything had to be approved by a historical commission. The exhibits were created by us. I formed a committee with a scientist, a contractor, a graphic artist, an electrician, and a plumber. I did this because I took bids from a national exhibit company and the price for the exhibits was so high it would have depleted the rest of my budget. The parking lot and trails system that I had budgeted for would not have been completed. Fortunately, my sub-contractors and scholars who were friends of mine donated their time to finish the exhibits.

Our first exhibit was a groundwater model, not one to take to meetings like in the previous chapter where you had to hand pump water through the model. We wanted an elaborate model that you could determine draught, normal and heaving precipitation times, and the effects on the system. The exhibiting company wanted one hundred and fifty thousand dollars to build one. My friend Dr. Charles Fitts with the University of Southern Maine and Walter Andersen, emeritus geologist for the state of Maine both volunteered their time to develop the hydrogeology of the model. My graphic artist designed the graphics for the model to display the groundwater system. A local sign company built the model, and my electrician and plumber designed the working parts of the model. Our finished exhibit measure stood six feet long by five feet high and was amazing. We did this project for twenty-two thousand dollars.

Our next project was working with the University of Maine and Dr. Paul Mayewski to build a climate exhibit. Dr. Mayewski designed the exhibit and once again my graphic artist, sign company, electrician, and my plumber built the exhibit. This was a win for everyone. The university received a twenty-five-thousand-dollar grant for the project. From the exhibit and talking with the Maine State Climatologist we learned that this historical plant sat on an esker system, and this area experienced a climate shift regularly. To help the university with their research we installed a weather station that fed data to Boulder, CO every 30 seconds.

Working with Walter Andersen the Maine state Geologist emeritus, we designed a glacial exhibit. This exhibit created a detailed description, with graphics of how Maine geology was formed during the last ice age, and how the glaciers receded. It spanned the time from roughly fifteen thousand years ago to the present day and was twenty feet long and four feet high.

Maine has experienced earthquakes, which is unusual because there is no fracking in Maine, and geologically Maine is in the middle of a tectonic plate, far from active faults. The state of Maine geologist Dr. Robert Marvinney director of Maine's Bureau of Geology says the phenomenon is from the last ice age. Earth's crust is springing back from the pressing weight of the massive ice sheet that has since melted (Marvinney, 2011).

Twenty-five thousand years ago, part of the massive Laurentide ice sheet flowed down from the northwest, ending on what is now the Georges Bank. The ice was a mile thick, two miles in some places. The weight of this glacier compacted the earth's crust down five hundred feet. It melted fourteen thousand years ago. The crust was so depressed that coastal Maine sea levels were two hundred and thirty feet higher than today. Three thousand years ago, sea levels were two hundred feet lower than today from the rapid rebounding of the glaciers (Marvinney, 2011).

The final geologic display was long-term monitoring needed to sustain a watershed. This display showed how important it is to place instrumentation in the right areas, collecting data and water quality. This exhibit displays a staff gauge for streams, 2-to-4-inch piezometers for groundwater dept, and precipitation data along with water quality. In addition, visitors could see how this data was inserted into a database and fed into a groundwater and surface model. By modeling the watershed, we could determine the short and long-term impacts on the watershed.

One exhibit that was very popular showcased the impact of water on health. We all should drink on average eight glasses of water daily. This is not tea, soda, and other drinks. We should all drink just water for our bodies to have optimum health. This exhibit showed all the body parts that benefit from pure water. Water keeps your brain healthy, your kidneys flushed out, your blood healthy and your muscles pliable. Water is essential to life for every living thing on the planet. We have no choice but to protect our water and make sure we practice sustainability in how we use our water resources. We cannot take this for granted. We may be the water planet, but we only have about one-point two percent of water for human use. The rest is in glaciers, ice caps, and permafrost. Only three percent of earth's water is freshwater. The rest is in our oceans.

THE SCIENTIST

One of the most prominent and oldest research institutes is the University of Maine Global Climate Change Institute. This institute was the first with a multi-interdisciplinary focus. They combine many academic units that create heighten creativity with their research. By integrating field, laboratory, and modeling activities to understand the physical, chemical, biological, and socio-cultural components of the climate system of past and present, they can better predict future climate changes and how they impact us globally. This institute's investigations span the last two million years. This timeframe represents multi-millennial to centennial-scale climate changes punctuated by abrupt (annual to decadal) shifts in climate. These understandings represent the full dynamic range of the planet's natural systems that are evolving dramatically and are influenced by human activity (University of Maine, 2021).

This institute started over forty years ago when Dr. Hans Born, started the institute in 1973. Under the advice of Senator Edmond Muskie, he worked to develop a multi-discipline research facility, encompassing geology, biology, archeology, chemistry, and the social sciences. The long-term success of this organization has been the multi-disciplinary approach to research. Their goal is to look at past and present environmental concerns to better understand the future.

The current director is Dr. Paul Mayewski, who is a world-renowned scientist who is often interviewed on 60 Minutes. He leads a group of leading scientists to uncover the science and help predict the future of our planet. His research is primarily in the Antarctic, drilling ice cores that are approximately two to five inches in diameter by a hundred to two hundred to ten thousand feet long to understand how the climate has changed over the years. According to Mayewski, these are the most robust records of the climate, going back season by season and now storm by storm (Mayewski & White, 2002).

By calibrating this date with modern sea ice, they can gain a perspective on the long-term climate of the planet. Currently, we only have records dating back one hundred years, but by studying the ice, we can go back to see how volcanos, the environment, and how man has impacted the planet. They have been able to determine how different things have come together to change the climate. One is that the earth's climate can change abruptly. The greenhouse gases, toxic metals, and dust we put into the atmosphere can alter the climate in one or two years. It could then stabilize but shift back in a few years. The bigger question Dr. Mayewski wants to answer is "what will happen in the future"? This would help us plan for our future. What to grow, where to build, and basically how we should live (University of Maine, 2021).

Dr. Daniel Sandweiss is an archeologist who studies how climate change in the past has affected cultural change. Then using this data help model and predict the future. By studying sea life and shells he wants to answer the question, "how did climate affect people in the past, and how did they react and survive"? One significant finding is how El Nino and El Nina have

Figure 1. Ice drilling, research huts, ice drill, and ice vault at the University of Maine Orono
(UMO, 2000).

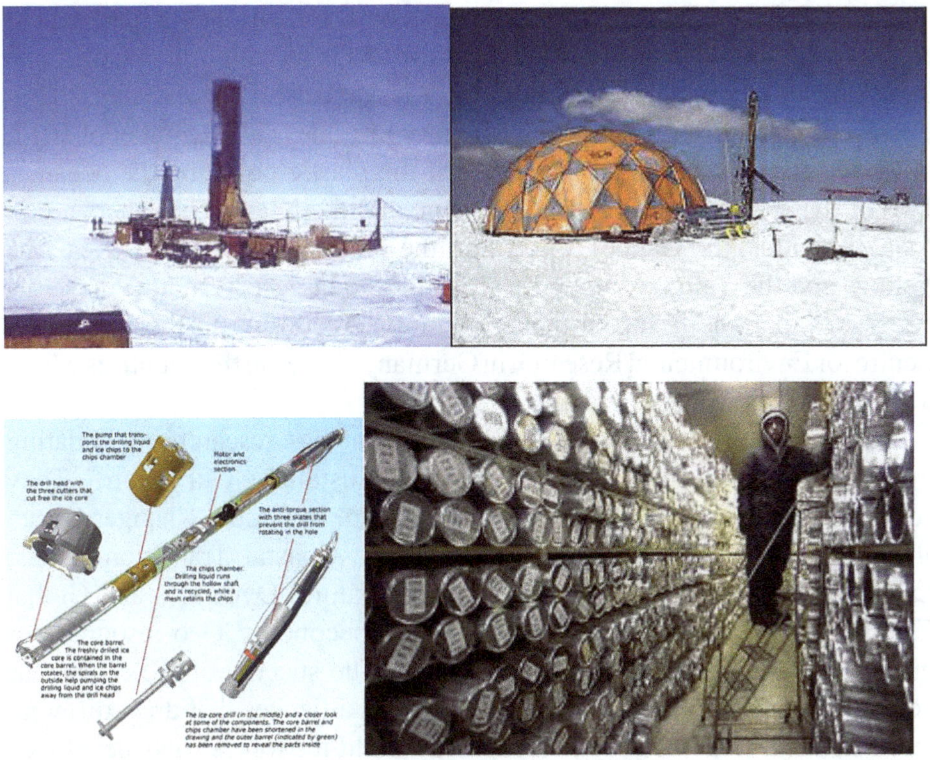

changed over the last ten to twelve thousand years. He discovered there are times when they did not exist, to times occurring more and sometimes less than today. This is important in determining how El Nino and El Nina could affect our future climate (University of Maine, 2021).

Dr. George Jacobson's, Emeritus Professor, focus is on how forests react to long-term climate change and the chemistry of the atmosphere, and the universal climate system. By uncovering the secrets of how the earthworks. Understanding the long-term viability of the planet, it helps to understand the modern world. Dr. Jacobson says, "Scientists are the new explorers, discovering things about the earth that no-one has ever known before" (University of Maine, 2021).

Dr. Stephen Norton, Emeritus Professor studies the chemicals found in the earth. By looking at lake sediment and peat cores, and by sharing this information with other colleagues around the world, models are being developed to predict what effect man's development will have long-term. Dt. Norton says, "going forward we have to determine how our behavior influences our environment, which is supposed to nourish and feed us". The long-term effects of atmospheric pollutants have a significant impact on our environment. Dr. Norton points out how the levels of pollution in China and India have led to an increase in pollution in Alaska (University of Maine, 2021).

This goes along with an article written by Sky News in December of 2017, "Just Ten Rivers Carry ninety percent of the plastic polluting the oceans". Two of these rivers are in Africa, the Nile and the Niger. The others are in Asia, the Indus, the Ganges, the Amur, the Mekong, the Pearl, Hai he, the Yellow, and the Yangtze. Not only are these the largest rivers they have the largest populations living in their proximity. According to the Helmholtz Centre for Environmental Research in Germany, these are the countries where waste control is lacking.

The Helmholtz Center for the Environment has research goals dating from 2007 to 2027 called "Changing Earth – Sustaining Our Future". They currently have nine research areas: Atmosphere in Global Change; Ocean and Cryosphere; Living on a Restless Earth, Coastal Transition Zones; Landscapes of the Future; Resources under Natural Dynamics and Global Change; Marine and Polar Life; Sustainable Bioeconomy; Geo-resources for the Energy Transition; One Healthy Planet. The success of the center has been the numerous cross-cutting activities and alliances that drew different research areas together promoted profile-building topics, and developed networks with other scientific partners in Germany (Helmholtz Center for the Environment, 2021). This alignment between non-profits, government agencies, and universities is a common theme among scientists.

One of the largest groups of scientists studying together and sharing this data is the "The University of the Artic (UArtic)". They are part of a network of universities, colleges, research institutes, and other organizations concerned with education and research in the arctic region. UArtic works to build and strengthen resources and infrastructure that better serve their regions. Part of a consortium of major universities globally, whose goals are to bring northern voices and knowledge to the global stage. Increasing understanding and respect toward the region, increasing human competence and capacity in the North, improving lives and communities for all northerners, and creating a healthy environment that serves the North and the world over generations (UArctic, 2021).

UArtic's research platform creates a space for sharing with the academic community, policymakers, and the public. These research projects and initiatives are developed by scientists from different disciplines and networks. Their focus on multidisciplinary studies has been successful in contributing to the Arctic's communities and ecosystems.

According to the Coastal Maine Futures Report, (2018), one of the most significant recent changes in climate relates to dramatic annual warming that has taken place on average across the Artic since 1980. This has been associated with a fifty percent reduction in the ocean covered by sea ice at the end of the summer melting. Arctic warming and associated sea-ice loss is thought to underlie an observed rise in extreme weather events across the Northern Hemisphere, owing to changes in atmospheric patterns resulting from what is now a smaller difference in temperature between the equator and pole (Birkel & Mayewski, 2018).

The southern hemisphere is being researched as well. The El Nino/Southern Oscillation (ENSO) plays an important part in global climate, with the exchange of heat and moisture across the equatorial Pacific Ocean. ENSOs are identified as phases that are determined by temporal changes, the warm phase El Nino is known to bring rising temperatures and rains on U.S. western coastal areas and causing significant weather issues elsewhere. La Nina brings the opposite, cooler temperatures and less storms. It is thought ENSO has the greatest effect on global weather patterns in the Northern Hemisphere winter, where impacts extend several months or gain in intensity. ENSO controls the Jetstream, and the strongest El Nino on the record developed in late 2015 and persisted into early 2016. The winter of 2015-2016 was the warmest winter on record for the Northeast. In 2016 the summer brought in the most significant drought in a decade. This may be a connection with El Nino (Birkel & Mayewski, 2018).

Many variables control climate. One of these variables is volcanic eruptions. When volcanos erupt, they spew material into the lower stratosphere, and this can cool the planet for years. This can happen regionally, hemispheric or on a global scale. This cooling develops directly from aerosol clouds that form and acts like a thin curtain, that keeps sunlight from reaching the earth's surface and can change the atmospheric circulation and oceanic response. In the Northern Hemisphere, volcanic eruptions tend to strengthen the winds across the subpolar and tropical North Atlantic Ocean. These increased wind speeds drive an oceanic response, where sea surface temperatures cool. In the Southern Hemisphere, there were eight major eruptions that occurred during the nineth and twentieth centuries. These were modeled with climate models, and it was determined that these eruptions significantly cooled the average temperature. A strong autumn and winter cooling response occurred following the major eruptions of Krakatoa, Santa Maria, Agung and Pinatubo. Understanding the impact that major volcanic eruptions can have on climates can establish early warning for food shortages and biosystems (Harvey, Grab, & Malherbe, 2020).

As we go forward, there are five major warnings considering climate change. First, is little or no increase in temperature, where the climate is trying to find out if there is a new normal. Most global climate models predict an increase in greenhouse gases, dust, and metal aerosols. Some areas show little to no increase in annual or seasonal temperature, whereas others have discernable increases. Extreme events that have come to mark the last decade with increased atmospheric blocking patterns that bring cold waves, heat waves, and heavy rainfalls have become what seems as a "new normal". Sea level increases are a part of a century-long trend, as the Greenland Ice Sheet continues to recede and grow as it tried to find an equilibrium (Mayewski, 2018).

The second is a moderate warming scenario, where an upturn in extreme weather patterns continues to increase in severity. The Arctic continues to warm, and summer sea ice continues to decline almost to collapse. The North Atlantic atmosphere patterns in the warm climate promote intensified influx of Gulf Stream waters, with rising seawater.

The third scenario assumes that all seasonal sea ice melts. This major collapse is driven by continued carbon dioxide emissions in addition to releases of methane. The Arctic Ocean is now ice-free from August until October or November. All atmospheric patterns are changed across the Northern Hemisphere every season of the year. The growing season in northern climates lengthens by two weeks, and winters bring more rain and ice, than snow. There is a reduced temperature gradient between the equator and the poles, which produces a slow-moving jet stream and more frequent weather events.

Fourth, major volcanic eruptions cool the climate for 1-3 years. If two or more major eruptions occur only years apart this can impact the climate for a decade or longer. Atmospheric circulation strengthens across the planet yielding several severe winters comparable to those of the past. It is possible to estimate volcanic activity by looking at past eruptions. However, nothing is certain (UAGS, 2021).

Five, determines that after several years of cool surface waters in the Pacific equator region, El Nino's shift from moderate to strong. The warmer conditions create drier areas and drought conditions for some and wetter areas for others (Burkel & Mayewski, 2018). Currently, NOAA predicts a normal ENSO for the year 2021. El Nino and La Nina are the warm and cool phases of a recurring pattern across the Pacific Ocean. El Nino's southern oscillation is known as ENSO. Researchers are now exploring if the sun controls this recurring pattern with solar storms. The pattern can shift back and forth irregularly every two to seven years. Each specific pattern can alter temperature, precipitation, and wind speeds. They can be large-scale weather events that trigger global events (NOAA, 2021).

The New Zealand Climate Change Research Institute (CCR) was established by the Victoria University of Wellington in 2008 with the goal to develop interdisciplinary research into all aspects of climate change. This small group of scholars and decision makers has worked together to develop research topics of real interest so that policymakers in New Zealand and elsewhere can better understand and make informed decisions regarding climate change.

These scholars have a particular interest in the natural and social sciences with an emphasis on working with international scholars. This draws on the skills of many different disciplines to create better governance, and to be able to anticipate, mitigate and adapt to climate change for both the private and public sectors.

A recent paper by Harrington and Frame (2021) shows extreme heat conditions are becoming more frequent for countries nearer the equator which generally are poor nations. This will affect so many issues, including water supplies. Their research determined that heat extremes do not increase evenly everywhere but is becoming more frequent in countries nearer the equator. The disparity between the global rich and poor will only continue to get worse.

The Climate Change Research Institute (CCRI) of New Deli, India was founded with a vision to develop and promote understanding of climate change, mentor both private and public sectors, and develop solutions for adaptation, mitigation, and human resource capabilities while addressing carbon footprints and emissions in the atmosphere. CCRI is a non-profit organization, with a specific mission to inform youth in schools and colleges about the environment, ecosystem changes, and consequences of climate change. They hope to build awareness of scientific and societal issues such as water, air pollution, and energy (CCRI, 2021).

CCRI took a new approach in 2021, by networking, building campaigns, lectures, and bulletins that hope to reach the general society. By promoting education and teaching communities the science of climate change they will reach a wide stratum of society. The goal is simple, to create awareness. This is so important because by reaching and educating the general population, true change can take place on a community level (CRRI, 2021).

The Global Climate Change Institute of Denmark is another leading institution studying global climate change. There is a worldwide initiative, engaging in climate projects globally. Long-term the hope is to be visible and operational in most countries globally, and work in local communities protecting everything from the rainforest, to re-establishing forest areas, developing, and implementing renewable energy sources and green initiatives. However, their main projects are protecting water resources and oceans.

One of the largest projects the Global Climate Change Institute of Denmark is currently working on is providing clean water for Central America. The project is focused on providing clean drinking water to households in the Republic of Honduras. This is being done by distributing water filters by Native Energy, a U.S.-based company, (CGI, 2021).

The "NativeEnergy Hydraid BioSand Water Filter Program Central America" aims to reduce wood fuel consumption used for domestic/non-residential water treatment by introducing zero energy BioSand water filter technology to replace the use of boiling. NativeEnergy, an American social enterprise manufactures and distributes the Hydraid BioSand water filter to provide households in the developing world with convenient and affordable access to safe drinking water. The project reduces GHG emissions associated with the consumption of biomass for wood fuel traditionally used to boil drinking water (CGI, 2021).

Education is paramount, due to the filter maintenance and support. The water filter consists of enclosed layers of sand, gravel, and a biological surface layer that absorbs pathogens to clean water. It is powered by gravity and has a lifetime of ten years (CGI, 2021). This is only one of many projects the institute is working on. Projects involve carbon footprints and emissions, planting forests, reducing deforestation, and renewable energy (CGI, 2021).

Arizona State University has taken a different and unique approach to tackling our environmental issues. They have found, like the other institutions, that an interdisciplinary approach is needed to develop sound science and long-term ecological results. A "Decision Theater Network" actively engages researchers and leaders to visualize solutions to complex problems. The network provides the latest research and expertise both nationally and internationally.

One unique part of this program is the "Spirituality and Sustainability Initiative (SSI). Founded in 2015. This initiative integrates the perspectives on sustainability and innovation with an awareness of the sources, dynamics, significance, and meaning of values and ethical convictions. Because the questions and commitments associated with spirituality are intrinsic to a holistic approach to sustainability SSI even though secular, set about defining spirituality broadly to include humanistic values and informal modes of spirituality as well as established religious traditions (SSI, 2020). This will help with the development of a theoretical framework for integrating social, ecological, and technological systems for facing climate uncertainty in a more holistic way (Arizona State University, 2021).

The Rocky Mountain Institute (RMI) has developed a five-year plan, called "Accelerating the Low Carbon Energy Transition". RMI has been a leader in the sustainable movement from the beginning. Amory Lovin, the founder, was a co-author of the groundbreaking book "Natural Capitalism". RMI has researched and worked as a non-profit in the energy and sustainability field helping businesses and governments transform how they build, operate, and sustain their operations. RMI's plan is to transform the project in markets and work with enough partners to reinvent energy systems. Their goal is to gather forces to create clean water and air, energy security, job growth, and sustainable economic development while slowing the effects of global climate change (RMI, 2021).

RMI seeks whole-system transitions. By building a practice with the best scientist, economists, and technologies for over thirty-five years they have completely changed how many entrepreneurs build and operate their businesses. They pride themselves on thought leadership, by being a trusted, independent source of technical expertise, research, and analysis. RMI crafts integrative transition scenarios, that build an economic business case that allows all stakeholders to see what is possible. They do this by co-developing solutions that will have a market impact. Shaping markets just sounds crazy, but by bringing stakeholders from all sides of the energy system, they hope to overcome market barriers and help change our energy system. Not by shutting it down but by transforming it over a period of time (RMI, 2021).

It is interesting that the scientific community is not divided. It is our media and politicians that are divided. This shared sense of purpose among so many scientists, non-profits, and communities are going forward without the ones yelling and screaming and ready to take down economies. The key is the word "communities" working with the scientist to bring about a better world and work toward the greater good. Science is ever-changing, sometimes very quickly. A major volcanic eruption can send us back to a mini-ice age, or a long El Nino can give us floods and droughts. Man can change behaviors, and they do that only with education. It is hard for people to know the truth with the mixed messages from our media and our politicians. It is more important

Figure 2. Rocky Mountain Institute Zero Net Energy (2020).

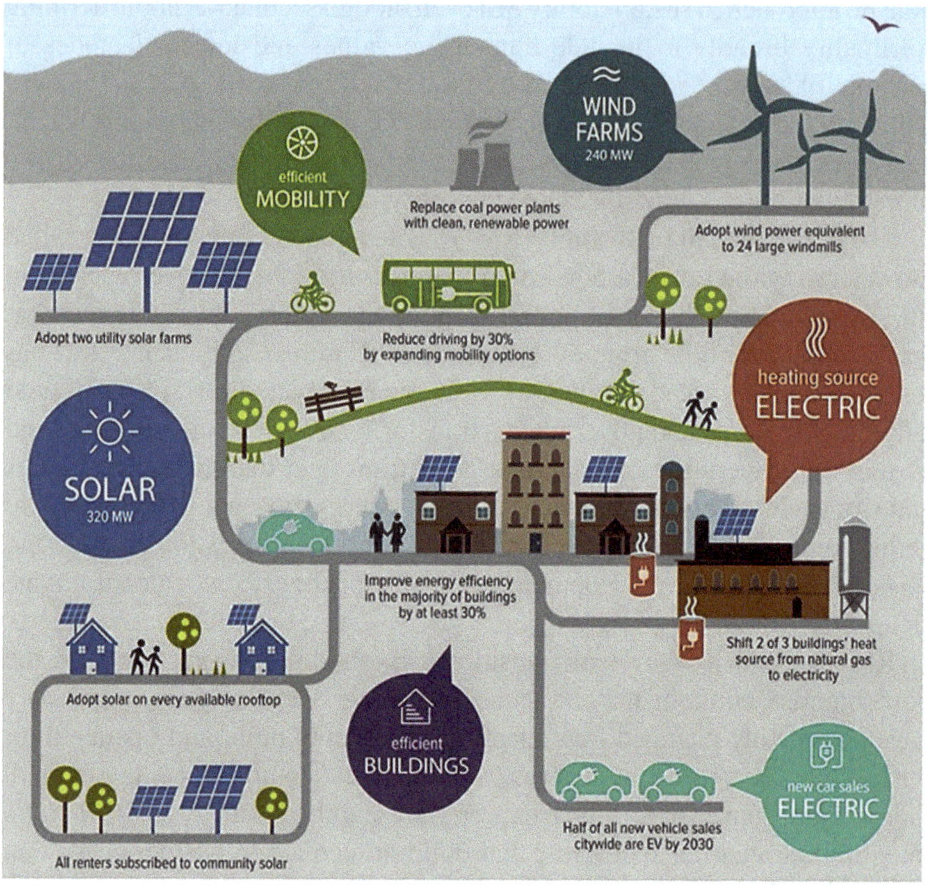

now than ever for education to teach science and not make science political. We will win sustainability with truth and reasoned debates. We will not win with ideology and political divide.

REFERENCES

Arizona State University. (2021). *Climate Change and Adaptation.* Arizona State University. https://sustainability-innovation.asu.edu/research/theme/climate-change-and-adaptation/

Birkel, S. D., & Mayewski, P. A. (2018). Coastal Maine Climate Futures. Climate Change Institute, University of Maine.

Climate Change Institute. (2021). *Research of the Climate Change Institute.* University of Maine Orono. https://climatechange.umaine.edu/

Climate Change Institute of New Delhi, India. (2021). *The vision of the Climate Change Institute.* Climate Change Institute of New Delhi, India. http://ccri.in/aims_objects.htm

Global Climate Institute of Denmark. (2021). *Research and Projects.* Global Climate Change Institute of Denmark. https://www.globalclimateinstitute.com/en/climate-projects/

Harrington, L., Frame, D., Fischer, E. M., Hawkins, E., Joshi, M., & Jones, C. D. (2016). Poorest countries experience earlier anthropogenic. *Environmental Research Letters, 11*(5), 055007. doi:10.1088/1748-9326/11/5/055007

Harvey, P.J., Grab, S.W., & Malherbe, J. (2020). *Major Volcanic Eruptions and Their Impacts on Southern Hemisphere Temperatures During the Late 19th and 20th Centuries, as Simulated by CMIP5 Models.* doi:10.1029/2020GL087792

Helmholtz Center for the Environment. (2021). *Research.* Helmholtz Center for the Environment. https://www.ufz.de/index.php?en=33573

Marvinny, R. (2011). *Glaciers of Maine.* The Maine Geologic Survey.

Mayewski, P.A., & White, F. (2002). *The Ice Chronicles.* University of New England Press.

New Zealand Climate Change Institute. (2021). *Research.* New England Climate Change Institute. https://www.wgtn.ac.nz/sgees/research-centres/ccri/research

NOAA. (2021). *ENSO.* NOAA. https://www.climate.gov/enso

Sky News. (2019). *Just 10 rivers carry 90% of the plastic polluting the oceans.* Sky News.

University of the Artic. (2021). *Research.* University of the Artic. https://www.uarctic.org/about-uarctic/

USGS. (2021). *Volcano Observatory.* USGS. https://www.usgs.gov/volcano

Chapter 11
The Squatters

ABSTRACT

Squatter's rights are laws that allow a squatter to use or inhabit another person's property in the event the lawful land owner does not evict or take action against the squatter. These laws usually only apply if an individual has been illegitimately occupying a space for a specific period of time. A squatter can prove adverse possession, in which they have a legal right to remain on the property. To prove adverse possession, the squatter must have taken possession of the property for an uninterrupted period of 12 years. Moreover, the legal owner must have abandoned the property. There are multiple ways to remove a squatter from a property, and each depends on the level of the squatter's resistance. You can have the police remove the squatter as they are a trespasser; apply a claim, a warrant of possession under the Residential Tenancies and Rooming Accommodation Act 2008; or apply to the Magistrates Court for a warrant of possession under the Property Law Act 1975. Through autoethnography, this chapter explores squatters.

If you give resources, then you must build capacity, or you will create generational welfare. – Ruby Payne

DOI: 10.4018/978-1-7998-9691-3.ch011

MEMOIR

It was summer, the best time to look for spring water. The ground is hot, the and springs are cold. A young geologist and I flew over an area in a helicopter and thought we saw some promising sites. We took a U.S. Geologic map and put grids on, an X and Y axis. I took the X and he took the Y. We took our equipment and started off. We had radios to communicate with each other, water, cameras, Garmin GPS meters to get coordinate if we found something good, and I packed lunch. Our plan was to check in frequently, leave our radio's on in case one of us got into trouble. We both had to bear spray, and a nine-millimeter handgun in case we ran across a rabid animal

Midway through the afternoon portion of the work, I came to a clearing and encountered two pit bulls and a guy skinning a deer. It was an old camp with junk around it. The pit bulls went nuts; the guy grabbed his shotgun. I put my hands up, and said, ‘I am leaving, everyone has a right to eat. I started backing away. Of course, Dave heard everything and started asking if I was okay. I reassured him I was, and that I would meet him back at the car. I stayed looking at the guy in the eyes, I informed him when I was at a good distance, I was turning around and he would not see me again.

As I was going through the woods to get back to our vehicle, I stopped hearing the dogs, but I knew he was following me. I guess he chained his dogs and wanted to make sure I left. He was tracking me! I met up with Dave at our vehicle and said just put everything in and we will pack it up later. We got in and drove off as quickly as we could. Finally, we hit a state highway, and I breathed a sigh of relief. Dave and I hadn’t said work until then.

We stopped and packed up, and had a beer to debrief. We were lucky! I think the guy skinning the dear heard the radio and knew I was not alone. He did not know how many of us there might be so he let me leave. I knew by his eyes and countenance that I had met a dark person. I knew he would have no qualms killing me. That was the first and only time in my life that this has happened to me. I pray it never does again for me and mine.

The funny thing is, this was not the first time we encountered a squatter. Most squatters are friendly, offer us a drink and actually help us get our bearings. Some have even come with us to help us avoid dangerous places. Most just ignore us. Squatters such a this, are just people who want to be alone, live off the grid, eat wild game and from their gardens. Surprisingly most that I have encountered are veterans, and their service has them burned out.

Figure 1. Squatters in a national forest
(U.S. Gov, 2022).

I never felt in danger or threatened until that day. The pit bulls alone would have killed me. I have nothing against pit bulls, just how they are raised. My daughter has one, he is the biggest baby I know, and loves all of us.

We decided to end our ground searches in the wild woods, and contacted NASA to see about renting a thermographic camera (when this took place, we did not have the modern technology we have today). We strapped the camera on to a helicopter. Then I strapped in to work the camera. The camera and I were hanging half out of the helicopter. We were targeting areas where the ground was significantly colder. This could be spring water. It was a blast! I had the best time. The ex-Navy Seal flying the helicopter loved it because I was not afraid. I wish I was young today, as opposed to back then. I would have had so much fun! This would have kept us off the long ground searches, and more easily targeted areas of interest.

WILDERNESS SQUATTERS

This week a big story coming out of New England is about 81-year-old, David Lidstone. David has lived in the woods of New Hampshire along the Merrimack River in a small cabin for the last thirty years. He has grown his food, have solar panels attached to his cabin, cuts his own firewood, and has chickens and a beloved cat. He is known by local kayakers and boaters as "River Dave". On July 15, the owner's son, having recently inherited the property, filed a civil contempt and had River Dave arrested. They confiscated his possessions and the rightful owner plans on demolishing his cabin. A support group of local kayakers has come to his defense helping him to collect the property taxes he owes for living on the property. They all love him!

According to a Merrimack County Superior Court Judge, he agrees that River Dave is not hurting anyone, but regrettably is on the wrong side of the law. New Hampshire's motto is "Live Free or Die", so there is plenty of sympathy for River Dave. The landowner has plans for the land, and did not know River Dave was living on the land. In addition, River Dave's cabin is in violation of local and state zoning, and environmental regulations and there is no access to a road.

The woodlot River Dave has called home is on seventy-three acres, that have been used for timber harvesting. The same family has owned the property since 1963. The owner is concerned about septic waste disposal, and solid waste. This woodlot was supposed to support the owner's elderly parent's retirement. River Dave does not blame the property owner and is accountable for his actions. A judge in 2017 asked Lidstone to leave, but he resisted. Now the town has hired a mediator. Currently, Lidstone can be released if he agrees to leave.

Lidstone, is a U.S. Air Force veteran, and father of four. The wooden, two-level A-frame cabin was profiled by a local television show in 2018. There is a small, cluttered kitchen with pots hanging from the ceiling, some appliances, and curtains on the windows. His porch has a footstool with a base made of stacked beer cans. He converted a wood stove into a beehive. He has attached lights, a mirror, and a pulley for a clothesline. Lidstone decided to live in the woods because it was the lifestyle he wanted. Currently, this time, his cabin has been burned to the ground, and an investigation has been started (Penn Live, 2021).

A common term for people like Lidstone is wilderness squatters. I came across many in my work. Some live on public land. They admit their lifestyle is unconventional. They want to live alone, with the beauty of the wilderness. They do not want others joining them. In the backcountry, if one person is living there, then another joins in and another, then there is no point living there. The squatters are people who do not like living in the domestic world. They think they are justified because they do not make as much of a mess. There is righteousness in a quiet sort of way.

These people live in shacks or huts. You may describe them as forest dwellers or homeless. One such individual, a man named Besh lived in a tree house on three thousand six hundred acres of mountain wilderness. Besh lived here with his wife. The San Mateo County sheriff found them and knocked down the tree-house which had been by them, and on the mountain for twenty years. The county was concerned for the couple's safety. Fires were the county's biggest concern. There was no running water, no sewage, and no facilities for this couple. There was a pit toilet, and students used to visit the tree house (Living on Earth, 2003).

Woodrat lives in a hut. He does not have much, but he does keep a nature diary. In his secret community, there are five people. All have their own huts. They try not to draw attention to themselves and are careful where they walk, so they do not leave trails. They draped military camouflage over their huts and hid the trails in tall bushes. Woodrat has a camp stove, a little propane heater, and a well-stocked kitchen. He goes to town, but not often. Even though this little community has to hide from law enforcement, they still like living in the woods.

Lenny has been Woodrat's neighbor for seventeen years. He is in his fifties and identifies as Jewish. He dresses in black with a pointed hat and looks like a hobo. Lenny says the squatters are modern-day Robin Hoods. The sheriff is the one wanting to kick them off the land. According to the Natural Resource Manager for the University of California at Santa Cruz, which owns three thousand acres in forested land, he estimates around a hundred people live in the forest. Some of them are students saving money and using the bathrooms at the school gym to shower. The non-students are choosing to be there for political reasons (Living on Earth, 2003).

There are several blogs to help people be wilderness squatters. One is called Rewild.com. Several people participate and offer free advice. Another is Squat the Planet. This site details step by step instructions on preparation, where to sleep, and how to live. They also open up a forum to questions and answers. There is a sense of romance to the idea of living in a quiet spot, in the woods, and building a cabin from local timber. Living off the grid, dodging rent, living the dream, and cozying up to the fire. That is not reality! There is a doctrine in law called "adverse possession", which we have mentioned before. This is the legal principle behind the so-called squatters' rights. The BC Limitation Act States that unless a right to land came into effect prior to July 1, 1975, "no right or title in or to land may be acquired by adverse possession. If you did not have your cabin built in the last century, you are going to have to show that you were squatting before 1975. Furthermore, the state can still shut you down, for health, sanitation, and fire concerns.

Some squatter's movements have a unique story. The Wilderness Center at Plump Hill in the Forest of Dean, in Gloucestershire County, was built to teach children and young people about the environment, nature, and the woods. As part of the public spending cuts, the county council intended to sell the center. Then the squatters showed up! Most are veterans, who say they have more specific goals. They still want a center to inspire young and old to create communities that live in a more sustainable way (Guardian, 2012). They are there to make sure the center stays open.

The squatters have invited specialists, such as horticulturists, farmers, blacksmiths, and basket-weavers to share and exchange skills with anyone who would like to learn them. They are doing this to slow or stop food insecurity in the UK. They want to teach others to grow and prepare their own food. Most of all to live in an environmentally sustainable way.

They see the center as a replica of an ancient Saxon house, a smoky, draughty building. By going back to their ancestors, they are ensuring the future. In addition, dreams of creating a community more in tune with the natural world to help people through mental illness. The reaction from the local community has been friendly, with farmers delivering supplies and good wishes (Guardian, 2012).

In Pombal, Brazil, the story is much different. For the past fifteen years, squatters have been cutting down the rainforest. The Brazilian authorities are thinking of giving the land to some squatters, so they stop cutting down the trees. They are hoping this will be the economic incentive to stop the squatters. Critics doubt this will stop them, stating they will continue to land grab. However, the government will be monitoring with satellite imagery. In addition, Brazil is so vast, it actually is a key player in regulating the planet's global temperatures. When there are no longer trees to absorb carbon from the air, there are no longer trees to release oxygen back into the air. Years ago, it was large corporations that caused the most damage, but now it is small settlers who are to blame (Trevisani and Forero, 2020).

THE GREAT EXODUS

From interviews with state agencies, talking to business developers, looking at census data and investigating what top business publications are writing about, the U.S. could be experiencing a paradigm shift. Many young professionals are leaving urban centers and moving to vibrant micropolitan areas. In the news, reporters are focusing on the decay of urban America. In the past, states have experienced the draining of talented youth in rural communities for better opportunities and jobs elsewhere (Tardif, 2021)

However, research shows that our youth may be returning to rural America for quality of life reasons, many sacrificing salaries for a better place to raise children. An important contributing factor is civic programs, arts, and entertainment that are now provided in rural communities due to the development of the creative economies. Regions once thought dormant are coming alive, cultures re-evaluated, communities working together, and most importantly disruptive leaders changing the way we think through entrepreneurism and creative business development on a large scale (Tardif, 2021).

Mass outward migration is happening in the states of New York, California, Maryland, and Illinois; primarily from their large urban centers. The New York City boroughs experiencing the most departures are Queens, the Bronx, and Brooklyn. In Queens 17, 959 residents left between July 2017 and July 2018. Brooklyn had 13, 555 departures, Manhattan dropped by 1,079, and the Bronx by 7,953. The combined total lost was 39,523 citizens. Census data shows many are going to the southern and western United States, mainly Texas, but not This outward migration has not stopped, and from 2017 to 2019 has continued at a much faster rate. Eight cities that are losing the most millennials are New York City, San Francisco, Los Angeles, San Diego, Miami, Houston, Las Vegas, and Chicago. California's inward migration is based on high incomes, graduate degrees, and education. The state's outward migration is based on high school education, no opportunities, expensive cost of living, no affordable housing, expensive higher education, and taxes. Many of these families have children and want a better quality of life. Those moving from urban centers are tired of filth and crime. They are unwilling to establish businesses in areas where they can afford the rent but cannot handle the existing deplorable conditions (Eber, 2019).

Wisconsin lost about 5,000 to 10,000 individuals each year from 2010 to 2016. New York, California, and Illinois are the three states with the largest outward migration in 2017, all losing more than 100,000 residents. Wyoming and Alaska, per capita, are the top two states in this ranking, with each of them losing over 1% of their population to other states (Crowne, 2019).

A century ago, people were moving to California from the Midwest. Now, most inward migration to California is from other countries such as Mexico, China, and Taiwan. In 2017, there were 11 million foreign-born Californians.

The state population is 39 million. California is fast becoming a minority-majority state with increasing Hispanic and Asian populations. According to the State of California, projections see a 21% growth rate in multiracial Californians by 2030. Departing minority Californians are younger, poorer, working-class families whose heads of households are 35 years and younger, and make less than $50,000 a year.

New York has had a net drop of 186,000 residents in 2018 alone. This trend has been ongoing since 2011. There are many factors contributing to this drop such as high taxes, high cost of living, poor job climate, miserable weather, and more recently decay and high crime. It is contrary to California where the outgoing citizens are older and more educated (Nichols and Patch, 2019).

Kansas is ranked fifth in outward migration, with 13,159 residents leaving in 2017. The biggest reason for this was unemployment. "We have to make sure we have access to skills training for people who live here and we have to become a more welcoming community to get people to move to southcentral Kansas," said Keith Lawing, executive director of the Workforce Alliance of South-Central Kansas. This exodus is now creating a labor shortage in the state. With a creative economy, disruptive leaders could be the savior of the Kansas community. Moreover, a collaboration of universities, government, and businesses is needed to make this happen.

For a state to be competitive, research has determined it must have low taxes, low-cost infrastructure, low cost for services and offer a better quality of life. When states seek to "create areas and clusters of economic activity, promote a community's unique identity, and advertise rural communities as appealing places to live conduct business and attract visitors", they create an environment that is attractive to young professionals who strive for a better quality of life. (National Governors Association, 2019, p. 16). A paradigm shift, where states that have invested in their culture, have created cultural districts that are effectively mapped, have invested in extensive tool kits and training and funded innovation are attracting young talented individuals. Many of these places are categorized as micropolitan areas.

California. According to census officials, the top ten urban cities are showing an outward migration (Nichols and Patch, 2019).

Urban to urban areas showing growth is predominately in Texas, Arizona, and Florida. Texas has the highest gain of 490,000 people at the end of 2015. Florida gained 366,000 people at the end of 2015. Georgia, Nevada, Colorado, Oregon, and Utah all experienced a significant increase in 2015. In 2015 states that lost population were Connecticut, Illinois, Maine, Mississippi, and New Mexico. Many of Oregon's inward migration came from the outward

Figure 2.

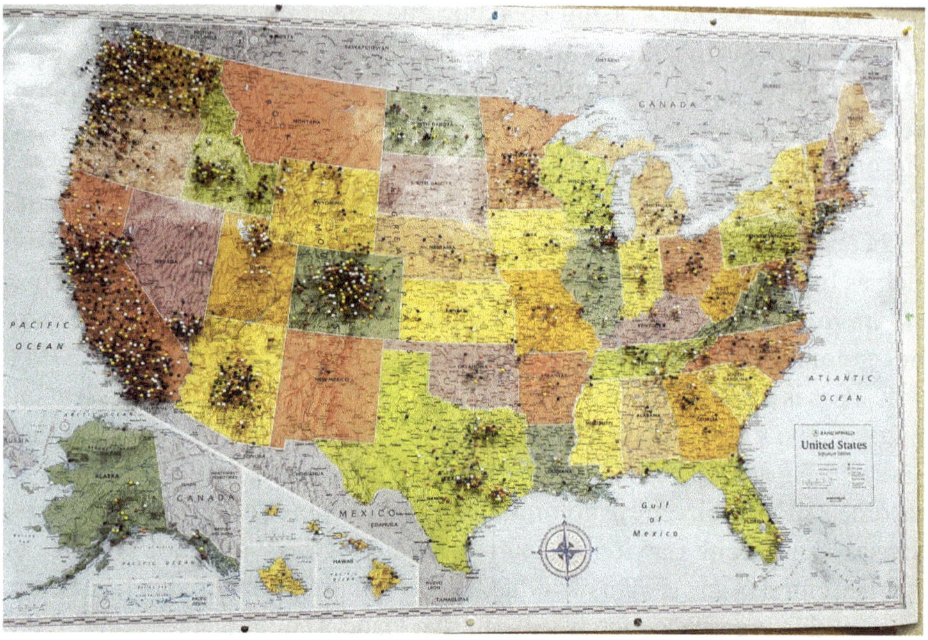

migration of California. Illinois, whose population dropped by 22,000 people to 12,859,995, is another state undergoing a similar trend. High taxes, lack of affordable housing, and quality of life are factors that led to the outward migration (Eber, 2019).

In Flathead County Montana, in the city of Kalispell, the town has had forty thousand residents move into the area, since 2016. They decided that each year, they would put a map of the United States on the wall, and put pins into it of where the people are moving from. Below is the 2021 map of households moving into the county (County of Fathead, 2021)

It is my hope that these people have joined the community, purchased houses and contributed to the local economy, and are not wilderness squatters. This is a wilderness area, so the impact to the environment would be significant. Any migration to an area impacts sustainability, and needs to have a long-term strategy in place so that significant impacts can be avoided. It is interesting that most of the areas people are moving from are urban. Flathead County needs to offer training to incoming households on the area and how to protect the environment and live sustainably.

In conclusion, whether it is wilderness squatters or mass outward migration, there will be issues regarding sustainability. The lack of sanitation and garbage removal, along with makeshift dwellings can be a health issue. In these wild areas, makeshift dwellings and equipment can lead to fires. Squatting communities jeopardize the management of our wilderness areas, where due to budget cuts may just have on forester for thousands of acres. For smaller areas, the impacts will be much greater (Ferguson, 1996).

Contemporary migration both nationally and internationally is challenging. There have been issues with agriculture, and short-term food supplies with migrations have been documented. Many challenges and opportunities often arise. Usually, an adjustment period takes place, learning a new culture, and often a new outlook on life. This is often healing and transformative. Being in a rural area gives one an opportunity to connect with nature.

Urban Squatters

Many of the outward migration groups from urban areas are leaving because they have a job in an area that will provide a better quality of life for their families. Some have learned (especially during COVID) that they can work from home; they have put their houses up for sale and in repressed areas this takes time. Others are leaving because they just cannot pay their bills where they live, and have taken employment in other areas where they are able to find work. Many have given their homes back to the banks.

These empty houses have been a target of urban squatters. Many of these squatters have named themselves the "urban homesteaders". This movement is not just in the U.S., but all over the world. Some of these groups, take care of the property, and some have actually been given the property by the government so that the property can be utilized efficiently. Not all property is treated this way. Most of the abandoned homes or houses for sale in states such as California have been destroyed by squatters. Trash, food, damaged kitchens, windows, and lawns have rendered the property unsalable. This individual squatter is not the same as a squatter settlement (Older, 2017).

Urban slums and squatter settlements are widely defined depending on a country's parameters. The general definition is: a residential area in an urban locality inhabited by the very poor who have no access to tenured land of their own, and hence "squat" on vacant land, either private or public. Millions of poor people have found moving into these areas, will improve their living and environment, and may even lead to a job. Until they find work and save money, they have to wait for housing.

A squatter settlement has no legal status, services, and infrastructure. These services would be a network of social infrastructure such as water supply, sanitation, electricity, roads and drainage, schools, health centers, marketplaces, and so on. There are many NGOs (non-governmental organizations) who provide help in these settlements. Squatter settlements have been in existence for centuries. After WWII, many came into existence because of the massive destruction that was left behind (Srinivas, 2015).

One of the best approaches has been to give or share buildings with the settlement and make it a state or government-owned facility where NGOs can help by providing social services to develop the capacity of its residents. Economies of scale in urban areas (large cities) will continue to attract squatter settlements. Instead of a confrontational approach, an enabling approach, where resources can be generated might be a solution (Srinivas, 2015).

REFERENCES

City of Kalispell, in the county of Flathead, Montana. (2021). *Map of inward migration*. Author.

Cornerstone Law Offices. (2021). *Property Squatters: How to get them Off?* Cornerstone Law Offices. https://www.cornerstonelawoffices.com.au/property-squattershow-to-get-them-off/

Ferguson, B. (1996). *The Environment Issues and Public Cost of Unguided and Information Settlement.* https://journals.sagepub.com/doi/pdf/10.1177/095624789600800218

Goldsmith, R. (2016). *Bursting the bubble: squatters rights.* Revelstoke. https://www.revelstokemountaineer.com/bursting-the-bubble-squatters-rights/

Living on Earth. (2003). *Wilderness Squatters.* Living on Earth. https://www.loe.org/shows/segments.html?programID=03-P13-00028&segmentID=10

Oyler, L. (2017). Urban Squatting's History is More Radical Than You Imagined. *New Republic*. https://newrepublic.com/article/142863/urban-squattings-history-radical-imagined

Penn Live. (2021). *Squatter David Lidstone, Fire Destroys cabin of 81-year-old man.* Penn Live. https://www.pennlive.com/nation-world/2021/08/fire-destroys-cabin-of-81-year-old-man-jailed-over-accusations-of-squatting-for-decades-in-woods.html

Srinivas, H. (2015). *Defining Squatter Settlements*. GDRC Research output E-036, Japan: Global Development Research Center. https://www.gdrc.org/uem/squatters/define-squatter.html

Trevisani, P., & Forero, J. (2020). Squatters Cut Down the Rainforest and Brazil Wants to Give them Land. *The Wall Street Journal*. https://www.wsj.com/articles/brazil-tries-novel-fix-for-amazon-deforestation-legalize-squatters-11580486956

United Nations. (2018). *Technical Report: Linkage of Migration and Food Insecurity, Agriculture, Sustainability to Rural Communities*. United Nations. https://www.fao.org/3/ca0922en/ca0922en.pdf

Chapter 12
The Class Action Lawsuit

ABSTRACT

Tort law was supposed to protect the innocent and be a means to retribution for those who were wronged. With all things, there are those who find ways to be greedy, and who do not care about ethics. It is those lawyers who give the profession a bad name. The author met a few of them, and its makes her sick just to write about them, and how they have negatively impacted innocent people. The tort definition refers to specific laws set out to determine whether a particular party is liable for harm caused to another party, and to determine the amount of compensation owed to the harmed party. Harm can be anything from injuries and pain and suffering to property damages and loss of wages. Through autoethnography, this chapter explores class action lawsuits.

Be true to yourself and you will never fall. – Beastie Boys

MEMOIR

On November 16, 2001, my world changed forever. My eighteen-year-old son died in a car accident and my family literally went into shock. I took the rest of the year off, and my family lived in numbness and limbo. I went back to work at the beginning of the New Year, and the first hour I was in my office a corporate accountant told me to get back to work and get over it. I

DOI: 10.4018/978-1-7998-9691-3.ch012

hung up the phone and wondered how someone could be so cold and cruel. I am a work alcoholic, always working full steam. I went back to work and shut everything out, the grief, pain, and shock. To make matters worse there were rumors we were about to face a class action lawsuit.

On January 28th I had members of our executive team at a spring site thirty miles from my office, which was located at the company museum. My staff called me and said that a small bus of suspicious men had trespassed on our spring property, and acted strangely in the museum. The executive vice president with me at the time told me she thought they were a group of torte lawyers who were planning a class action lawsuit against the company. We all piled into my car and headed back to my office. As I drove, my staff updated me. I called the county sheriff and informed him I had pictures of the group trespassing on our property, and security cameras had caught one of them taking a plant tour video out of a museum exhibit and copying it on his laptop. This video was proprietary and copyrighted.

I had the group arrested and found out the main lawyer was a famous tort lawyer, who was in partnership with two other tort lawyers; the kind that flies around in their corporate jets. So began a journey where I observed the best and worst of lawyers, people, and my own company's executive team. I found myself in the ultimate conundrum. I was forced to keep my balance when constant pressures met me in every direction; even at home. I was moving through a multidimensional space of tightropes. I had to find ways to influence the media, lawyers, government officials, and the public. All the while trying to deal with my executive team, and all the conundrums that came from trying to keep them content with the work I was doing. The company made me the point person for the lawsuit. I was to handle everything on the ground. This meant I was on the video release to inform the public, deposed by twelve lawyers, met with government officials, and interviewed by thirty-five media outlets, which included all of the top national television, radio, paper, and magazines. I had to act decisively in a world of ambiguity. This ended up being a three hundred-million-dollar class action lawsuit, with a two-billion-dollar brand and twenty-five thousand jobs at stake. I received death threats daily,

This group of tort lawyers made their money by suing large corporations for a big payout. To them, it did not matter if they were wrong, the truth did not matter. They would spin a story for a big payout. They all used the same plaintiff for every case. I'm sure they made good money too, because they were ruthless, and would do anything to win.

This all came about because of quality control and meeting the demands of products being on the grocery store shelf. I talk about this extensively in the quality control chapter, so I will not go in-depth here. FDA defines spring as coming to the surface on its own pressure. To meet HACCP (Hazard Analysis Critical Control Point) compliance, you have to show that you comply with seven steps. These are the food-grade materials you use, sanitation, and testing. When you let spring, water come into a catchment, that catchment must meet the HACCP materials list, and be regularly sanitized and tested to be free of any contamination. Including air-born bacteria. In 1995, a revision was made to the bottled water laws, stating that a borehole (well) can be drilled in the vicinity of the spring, and it has to correlate to the spring. If you pump the well it should register in the spring, but you cannot dry the spring up. In addition, the water chemistry has to match. The state departments of natural resources will determine how much water can be removed, and approve the long-term monitoring of the site.

At the time I had several spring owners that wanted us to purchase their property. Also, at this time we were building manufacturing plants with spring developments across the country. So, this was not a priority. However, I would often meet with these spring owners. If we changed our equipment, I would sell the old equipment at a low cost to them. I had a great relationship with all of them with the exception of one. This one exception actually made us hesitant to purchase the other springs without long-term monitoring.

Before I worked in natural resources, I was a quality manager. Another person made a contract with an out-of-state spring owner. He was highly educated and had a large parcel of land with a spring on it. We started using his spring water and testing it frequently. We found problems with water quality. There were periods when the water had high bacteria counts and coliform. I went to check this out and started visiting with neighbors and city officials. There was a dam upstream, that released water when it rained, and this was causing the issue. I could not change this, so we quit using the spring. The owner went nuts. Remember Ted Kaczynski, the Unabomber was highly educated and was a mathematics prodigy.

I found out from the city and neighbors, that this guy even though highly educated had issues. He wanted to build his own community on his land, and have this community be self-sufficient. The neighbors and the city stopped this from happening. So, he sued everyone and would show up on their property yelling about injustice. When we stopped using his water, he did the same to us. He would show up unannounced and start yelling in our office about the injustice of the situation, and how dare we stop using his water. He was unglued! He scared our office staff, threatened me and my boss, and we had to have the local sheriff escort him off the premise. A highly educated nut case!

He hired one of the biggest tort lawyers in Boston, and he called all the spring owners who wanted us to purchase their spring and represented them in a class action lawsuit. His lawsuit stated we were not spring water because we collected the water in a borehole. In addition, he said our water quality was bad. The water quality data sheets he had were from the spring owner we stopped using. Which we proved to him in discovery. His big case was unraveling. He is the one I caught on camera stealing a plant tour video, by pulling it out of a machine, and taking it into the bathroom with his computer. A camera caught him coming out of the bathroom and putting it back into the machine. I am sure a jury or judge would come to the same conclusion.

When you get in bed with the devil, you might have more than you bargained for, and that is what happened to this lawyer. He partnered with two of the biggest tort lawyers in the U.S. before he did his discovery. He was ready to negotiate and settle, knowing he had no case. The other two broke from him, and it was nasty. At this point, large companies usually settle to keep their name out of the press. We agreed to purchase two of the springs, but not the crazy guy's spring or property. The are some lines you just have to draw. The other tort lawyers were not as ethical and wanted a three hundred million payout.

The lunacy started. We started a website to tell our story, national media came to interview me, and then the depositions. Along with all of this were the death threats. I would come to work in the morning and property would be dug up in the middle of the night to try to trace pipes. For a week solid, I had the sheriff out daily. We both just shook our heads in disbelief.

In this process, I found some interesting information about tort lawyers. They all have their favorite plaintiff. One group uses an ex-cop, one a teacher, another a banker. They would go on about how they were injured and the suit was valid. At the same time, I had just read John Grisham's book "The King of Torts". I thought to myself, "I am living this". It was a total circus. After we settled with the first tort lawyer, we took his advice and settled the overall case with a second one out of Chicago, this would end the battle with five others. The main lawyer I worked with looked just like Marilyn Monroe. I had a feeling it was very deceptive, and she would take no prisoners.

After three years of not being home, ten depositions, thirty-five national media outlets, state and local media outlets, and talking to every group imaginable, it was over. Now I had to decide on handling what we purchased, and the friendships were ruined in the process. It probably would still have happened down the road, with friendships intact forever. All because of a mentally ill man and the greed of tort lawyers.

Sadly, a corrupt legal system goes against everything we should be standing for in this country. U.S. judges who settle Tort law cases wield power. If on the take, they can make millions. Even if they are ethical; it can take years to solve a case. For companies, their reputations can be ruined by the length of the discovery and ongoing case. If the case is valid, a plaintiff could pass away before the case is finished. Sadly, the only ones profiting from this are lawyers.

Max Kennerly, Esq. has written about this in his article titled, "How Judges Can Settle Mass Tort Cases", (2012). In his article, Kennerly describes how judges need to be fair in high-stakes litigation. He points out that a judge hearing the litigation of the 9/11 workers believed the leverage sought by the parties was greater due to the supposedly sheer numbers, compared to the leverage defendants' have by virtue of their superior resources. Due to this the judge was willing to draw out the discovery.

In this case, the government had the resources to bring the litigation forward. As proved the court swiftly determined over forty percent of the plaintiffs did not even claim to have suffered a serious injury. Even serious cases, involving death, cancer, or heart attacks, were "likely to be vulnerable". After seven years, the case was settled just two months before trial. The case was solved without the judge, and negotiations between parties settled the case out of court.

TORT CLASSIFICATIONS

- *Intentional Torts* – occurs when an individual intentionally commits a wrongful act causing harm to another individual. The key distinction with intentional torts is the requirement of demonstrating that the defendant committed the tort on purpose.
- *Negligence Torts* – if the failure to exercise a reasonable standard of care, may result in a negligent tort. A negligent tort case to be successful, there need to be four elements present: duty, breach of duty, causation, and damages. Negligence torts make up the majority of tort lawsuits.
- *Strict Liability Torts* – these are strict liability tort cases, the focus shifts from the individual or party committing the tort to the act or

incident itself. In other words, strict liability cases do not take into consideration intent or even negligence on the part of the wrongdoer. What matters is the action and the resulting damages that occurred.

Tort Examples

Intentional Torts

- Battery – making unwanted, offensive, or harmful contact with another person
- Assault – an attempt to cause harm or an action of a threatening nature
- Fraud – an act of lying or making misrepresentations to another person
- Trespass – using another person's property without their permission
- Infliction of emotional distress – intentionally frightening another individual, causing severe emotional or mental distress

Negligence Torts

- Car/bicycle/motorcycle accidents - driver or rider speeds and hits a pedestrian causing injuries, the driver can be found negligent
- Slips and falls – claims when an individual falls and injures themselves on the premises of another person's property
- Medical malpractice – occurs when a medical professional does not provide a reasonable or competent degree of care, resulting in harm to the patient

Strict Liability

- Product liability – deals with cases of harm by defective products made available to the public by manufacturers, suppliers, or retailers
- Owning wild animals – owners of wild animals will be held liable should any of the animals escape and injure another person
- Exceptionally dangerous activities – an example would be a truck carrying hazardous materials leaking and causes harm because of the spill

The law was to compensate victims for damages, losses, injuries or pain suffered. The aim of the law is to restore the victim to the state they were in prior to the tort incident and to ensure the defendant forfeits any unlawfully obtained gains. When monetary compensation may not adequately address the impacts, a judge may impose equitable remedies.

The below memoir is an example of this law run amok and how greedy and unethical people can take something meant to help others and actually work to do harm all to make quick and high payouts.

Before 2017, Americans were reputed to be the most litigious society in the world, something happened, call it a tipping point, and this changed. Fewer than two in a thousand people filed tort lawsuits in 2015. Several factors have contributed to this, including state restrictions on litigation, the increasing cost of bringing suits, improved auto safety, and a long campaign by businesses to turn public opinion against plaintiffs and their lawyers. The massive number of lawsuits has confounded the public and their perception of tort lawyers. Companies on the receiving end of such lawsuits are relieved by the decline. Trade groups that represent companies have pushed to raise the bar for filing lawsuits and rein in damages, portraying a large chunk of tort litigation as a burden on judicial resources (Palazzolo, 2017).

Some judges are worried that the decline in the number of tort filings could mean that Americans no longer see courts as an affordable way to seek redress for their injuries. The reason is quite simple, it is expensive and time-consuming. Most tort lawsuits see damages for negligence rather than deliberate injury and fall into three categories: Auto Cases, Medical Malpractice, or Product liability. Even with the number down, tort lawsuits account for five percent of all civil filings in state courts. In addition, an election night poll by Public Opinion Strategies showed that eighty-seven percent of voters agreed that there are too many lawsuits filed in America (Palazzolo, 2017).

A recent study by the National Center for State Courts found that nearly two-thirds of tort cases involved automobiles, death, or injury from hospitalization with crash injuries. In addition, there was a fifty-seven percent decline in malpractice claims by doctor's insurance carriers. That could be due to the fact that thirty states have capped damages in medical malpractice cases (Palazzolo, 2017).

In Kansas, tort litigation moved to curb tort litigation. And tort filings fell by forty-five percent between 2000 and 2015. Kansas plaintiffs have to request permission from a judge to file for punitive damages. Many of these tort cases are solved outside of court and through arbitration. Since the 1994 case where a seventy-nine-year-old woman who sued McDonald's after spilling coffee in her lap and burning herself badly, even when the verdict was reduced and settled out of court, it was still considered a "frivolous" lawsuit by the public (Palazzolo, 2017). In my own case, I know the class action case against my company was a frivolous case, brought on by attorney greed. There is no doubt in my mind.

Recently, the conference of Chief Justices concluded that the state judicial systems are getting fewer tort cases because many disputes cost more to litigate than they are worth. In addition, lawyers make more money than the plaintiffs.

REFERENCES

FDA. (2020). *Rev. CFR Code of Federal Regulations Title 21*. https://www.accessdata.fda.gov/scripts/cdrh/cfdocs/cfcfr/CFRSearch.cfm?fr=165.110&SearchTerm=bottled%20water

Kennerly, M. S. (2012). *How Judges Can Settle Mass Tort Cases* (A Lesson from the 9/11 Responders Litigation). The Law Blog. https://www.litigationandtrial.com/2012/11/articles/series/special-comment/settle-mass-torts-cases/

Palazzolo, J. (2017). We won't see you in court: The era of Tort Lawsuits in Waning. *The Wall Street Journal*. https://www.wsj.com/articles/we-wont-see-you-in-court-the-era-of-tort-lawsuits-is-waning-1500930572

Valiente Mott. (2021). *What is Torte Law?* Valiente Mott. https://valientemott.com/blog/blog-what-is-tort-law/

Chapter 13
The Creative Economy and Sustainability

ABSTRACT

Creativity is fascinating. It is the result of mental activity, an action that occurs inside the heads of some special people. That statement may be misleading. If creativity is an idea or action that is new and valuable, then we simply cannot accept one's own account for its existence. Maybe creativity does not happen inside people's heads, but is an interaction between a person's thoughts and the social context they are in. Mihaly Csikszentmihalyi, the author of Flow, defines flow as working with passion and skill to achieve a purpose or goal. He determined there is a delicate balance between working in the flow and happiness. If you have the passion without the skill and experience one will have anxiety. The amount of anxiety will be determined by one's skill and experiences. If you have passion, skill, and experience, then working in flow gives you happiness and great accomplishment. Through autoethnography, this chapter explores the creative economy and sustainability.

At the heart of the creative economy are the cultural and creative industries that lie at the crossroads of technology, business, arts, and culture. – Unknown

Sudden bursts of creativity have led to the "Age of Discovery", "Age of Industrialization", "Age of Knowledge" and now the "Age of Artificial

DOI: 10.4018/978-1-7998-9691-3.ch013

Figure 1.

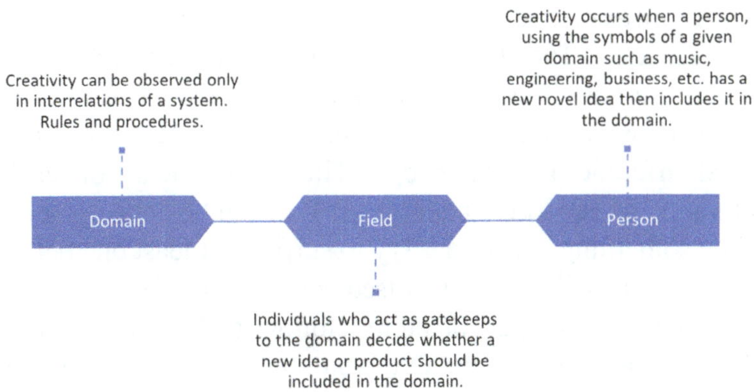

Intelligence". When one is working in the flow, time does not exist, they disregard their environment and give their surroundings a personal pattern that echoes the rhythm of their thoughts and habits of action. When the environment is their own making, they forget the rest of the world and concentrate on pursuing their muse. Today, centers of learning and commerce have been magnets for ambitious people who want to leave their mark on humanity (Csikszentmihalyi, 1996).

A person must be in a position to access the domain in which one plans to work. The information must be distributed equally, and if you want to learn something specific, you need to be in the learning environment created. These learning environments may have greater interactions and tend to lead to more experimentation. Access to the field may not be evenly spaced, meaning centers that facilitate novel ideas are not necessarily the ones where the information is stored or where stimulation is greatest. The ability to fund and act drives a change in the environment.

Creating a meaningful environment in space and time helps you become personally creative. When you do, life is rarely boring, and rarely out of control. Your life reflects you individually. You probably will not be recognized as highly significant or famous, but your life may be one of quality and a life well lived

Supporting families can positively shape creative lives. To build creative lives you must manage your time. By turning off the TV, and games, you focus on your creative environment, and children especially will develop their creative talents. Furthermore, research has determined that as people age, their creative processes increase as long as a positive attitude is maintained. An aging person with a positive attitude is more likely to forgive, not take offense, and have less anxiety. They have found meaning! (Csikszentmihalyi, 1996).

Our brains are not different, but how we think and what we think about are different. How we use our own mental energy creatively, is the most fundamental difference between people. The first step to a more creative life is the cultivation of curiosity and interest. This can be done by: trying to be surprised by something every day. Try to surprise at least one person every day. Write down each day what surprised you and how you surprised others. When something strikes a spark of interest, follow it (Csikszentmihalyi, 1996).

Mostly, to create flow every day, you start by waking up to specific goals. These have to be enjoyable to develop complexity. Develop habits of strength, when creative energy is awakened, keep it and value it. Manage your space and time. Change the way you think about yourself; this is psychic energy and can be positive or negative. A shift from closure to openness, aim for complexity. To realize one's potential, openness and complexity add a deeper layer of understanding, this allows one to solve complex problems (Csikszentmihalyi, 1996). Sustainability is a complex problem!

CREATIVE ECONOMY

Currently, creative economies are described in a variety of ways, depending on the composition and character of businesses, nonprofits, individuals, and venues that exist in any given area. The creative industries and the creative class have broad definitions; however, they contribute directly to jobs, tax generation, and wealth for the local economies they operate within. States have been studying the creative economy by looking at the work of nonprofits and cultural tourism. This may not be totally accurate; however, the creative economy encompasses technology as well (Tardif, 2020).

The majority of the creative economies are based on culture and cultural districts. These have a decentralized nature that can benefit residents of areas often thought to lack economic strength—such as rural areas and the urban core. At the heart of the creative industries are individual artists and micro businesses who are typically well-connected to the communities where they reside. Linking these artists and technology experts with entrepreneurial opportunities both inside and beyond their regions offers many economic development possibilities.

Figure 2. Linkages with cultural
(Tardif, 2020)

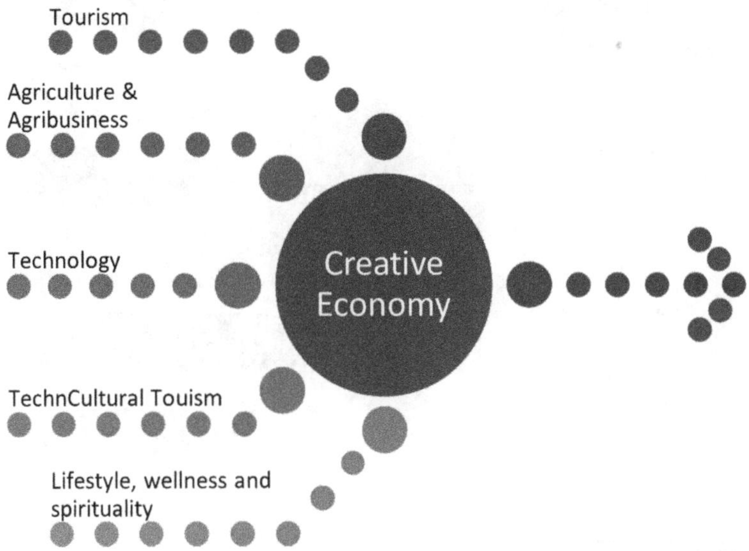

A state's arts and cultural resources can be economic assets. The arts and cultural industries provide jobs, attract investments, and stimulate local economies through events, tourism, consumer purchases, and tax revenue. Perhaps more significantly, new technology creates opportunities for a contemporary workforce and creates an appeal to a young skilled workforce.

Creative leadership and the creative economy not only have the potential to offer numerous benefits to socio-economic health but build a sense of place which builds a thriving community. Through the creative industries, states have an opportunity to create jobs, attract investments, generate tax revenues, and stimulate local economies through tourism and consumer purchases. In addition, creative industries are contributing to the contemporary workforce, making creative contributions to industries' products and services, and infusing culture into community development. The diagram below is a theoretical concept of core characteristics. The modes of developing the creative economy are through collaborative platforms with cluster developments such as micro business development, incubators with toolkits, cyber laboratories, new technology development, and creative maker spaces (Tardif, 2020).

Figure 3.

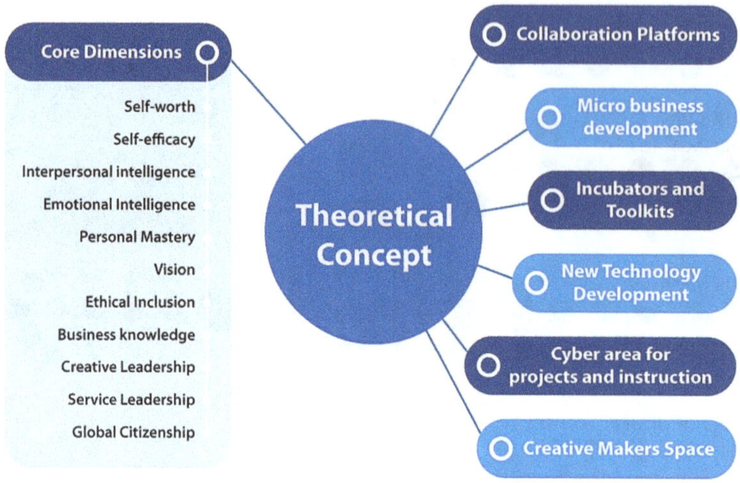

SUSTAINABILITY

There is no clear definition of sustainability. There are many concepts and viewpoints. Overall, it is the ability to maintain or a process that keeps things at a certain level that provides an environmental and social level for quality of life for living things on the planet. This would be done by being responsible for our natural resources so as not to deplete them. Being responsible for humanity and all living things. To live in balanced ecosystems for quality of life.

The role of creativity and the creative economy is a significant part of being sustainable. Creativity and sustainability could be the most important developments for mankind's future. Integrations of both these concepts could solve the complex problems regarding the protection of natural resources, by developing new technologies for energy, and the socio-economic issues that arise from change.

According to, Vivian Cheng's (2018), self-determination theory (SDT), there are innate psychological needs that motivate one's behaviors including the need for competence, autonomy, and social relatedness. An individual's behavior is considered self-motivated and self-determined. Personal interest, enjoyment, and inherent satisfaction regulate one's motivation. These intrinsic motives play a role in one's life-long behaviors. This approach coincides with Maslow's self-actualization theory of human creativity (1968). In this theory, creativity is considered a self-actualizing process. The creative process will bring about optimal happiness. Furthermore, the experience of engaging in creative activities or creativity training was found to have positive impacts on working, learning, motivation, confidence, and resilience (Cheng, 2018).

Modern concepts of sustainability recognize the many human interactions and the importance of creativity to sustainability. That creativity is a force to sustainability through its capacity for imagining and visioning and heightened creative problem solving (Albrecht, 2010). Thus, the need for more creative education may be the driving force to sustainability. Creative problem-solving opens pathways for breakthrough solutions to sustainability problems. With creative insight focus, not just a novel idea, but the creator of the emotional, cognitive, and psychological processes that will highlight key characteristics with more insightful problem-solving.

According to Mitchell and Walinga (2017), academic research and business practice have asserted that sustainability requires new business models and logic. The deterioration of our natural resources, economic growth, and economic inequity need ambitious sustainable development strategies. Transforming systems of production, logistics, and consumption requires the creation of different thought processes, without hurting the economic systems and the socio-economic conditions.

These changes need to begin with changes in thinking. Sustainability requires we move forward by minimizing the negative impact of production, and think more about recyclable processes, and reducing our footprints. This emergence of sustainability presents many challenges and barriers to transitioning to sustainable practices. Academics and practitioners are increasingly interested in how creativity can be applied to meet the challenges of sustainability (Mitchell and Walinga, 2017). Currently, business and management, sustainability literature has focused primarily on the role of leadership in stimulating employee-led eco-innovation and environmental performance, strategic collaboration, corporate accountability, and governance. These approaches advance the understanding of sustainability, they may not change deep-seated beliefs in the necessity for maintaining unsustainable production and patterns of consumption (Mitchell and Walinga, 2017).

This may lead businesses to reinvent themselves, develop new products and services that generate alternative conceptual models, and utilize applied tools that recognize the social and ecological impacts of sustainability. Mitchell and Walinga (2017), determined that creativity is at the heart of this collective transformation. Creative problem-solving characteristics, such as divergence and convergence, are essential at all levels of the organization as it creates sustainable futures. They proposed that the cognitive shift that sparks a transition toward sustainability happens not at the solution end of the process, but at the problem end of the process. They believed that insight results from a restructuring of the problem, which leads to an abrupt change in the solution pathway and leads the solver to immediate success with a problem.

ORGANIZATIONAL SUPPORT

To get to this much deeper level of problem-solving, where we are actually restructuring the problem, we need organizational support not only from the top down but from the bottom up. Also, achieving long-term sustainability will require innovations across all levels. Pro-active companies see sustainability as a competitive advantage. Many have adopted policies of sustainable development and environmental protection to attempt to decrease the impact on the natural environment. Even with this, there is still a gap between environmental policies and practices (Ramus, 2001).

Creative problem solving and innovative products and service ideas from the workforce can fill the gap between organizational statements and business practices. Innovation is the heart of the process. Actions taken by individuals or teams that improve the environmental performance of companies is at the heart of the overall mission, vision, values, and beliefs that make up the culture of a company.

According to Ramus (2001), half of the employees surveyed in environmental proactive companies have ever proposed a solution to an environmental problem. Those that did, felt that their direct supervisors encouraged environmental actions. Innovations are prominent in organizations today; they occur in companies encouraging creative ideas from all employees at every level in the company. These have come from three basic practices:

- Decreased the environmental impacts of the company such as recycling and reuse
- Solved an environmental problem for the company (like hazardous substances reduction)
- Developed a more eco-efficient product/service such as a resource or energy use

Organizational culture plays a role in generating ideas. It's part of their mission, vision, and values. In addition, many companies provide incentives, however, this needs to be worthy of the result. If an employee's idea saves thousands of dollars each year on energy costs, the incentive should reflect the savings. Major savings, the incentive should be significant, minor savings the incentive should reflect the minor savings.

Direct supervisory support is essential to success. Ramus (2001) interviewed fifty employees from five firms with a proven record of encouraging employee environmental initiatives. Employees said that supervisory behaviors encouraged them to support environmental innovation. Also, competence building, communication, information dissemination, and the use of incentives with recognition led employees to engage willingly to find innovative changes and new ideas. Ramus (200!) found six categories of supervisory behaviors that contributed to employee engagement:

- Encourage new ideas
- Gave feedback and ideas whether they were adopted or not
- Supported experimentation with new methods and ways of doing things on a small scale, to discover if they were feasible and profitable
- Open to proposals for changes in the way work is done, especially in the environmental area
- Sends employees on job rotations and visits inside and outside of the company to learn about environmental solutions
- Discuss mistakes so that their team can learn for future endeavors

In addition, it is important to commit resources and time for employees to attend training, make sure employees have the competencies needed to do their job, and spend time developing a lifelong learning plan (Ramus, 2001).

Communication in all organizations is always perceived as lacking, but it does not mean it should be ignored. Good communication removes barriers, allows for open dialogue, and is a time for leaders to listen (really listen), be honest, and for leadership to accept criticism without being defensive. Communicate, listen and inquire! Not talk about a problem, own up to any issues, and discuss them. It starts at the top of the organization. If the top leaders are on board, the line manager will be on board. If an organization's culture is reflective, then everyone can have fun being creative.

MICROPOLITAN AREAS

In this current climate of outward migration from urban America to rural America, the landscape of rural America is changing drastically. Micropolitan areas now number over five hundred and growing. These are areas greater than ten thousand people and less than fifty thousand. They can be areas within an urban area, most have developed in rural areas. Many of these have developed thriving creative economies, with many sustainable innovations.

In the *Micropolitan Success Stories from the Heartland*, Ross DeVol and Shelly Wisecarver (2018) focused on five micropolitan success stories. According to DeVol and Wisecarver, several smaller communities (micropolitan areas) in the Heartland have out-performed other micropolitan areas on the east and west coasts and the nation (DeVol and Wisecarver, 2018). In addition, interviews with the micropolitan officials gave insight into a new economy to be developed across the nation. The five micropolitan areas featured in DeVol and Wisecarver's report are Findlay, Ohio; Brookings, South Dakota; Oxford, Mississippi; Jasper, Indiana and Ardmore, Oklahoma. These five areas were chosen because of their consistent growth in per capita income, wages, and jobs. Furthermore, all five micropolitan areas made "best lists" for economic growth and business expansion. Most importantly, all five have based economic development on the creative economy. They are not soliciting large industries, but rather small to medium creative and innovative businesses.

By focusing on the creative economy and innovation, talented young professionals are attracted to these areas. In addition, investment capital is also attracted to these areas. By pooling the arts and technology sectors, they are creating a vibrant economy and sense of place. This concept has given way to organically formed clusters. When one high-tech company comes to an area, it leads others to come. This forms a cluster unique to the area, which attracts highly skilled young professionals, who are not just there for the career options, but also for the art venues, restaurants, and amenities that are developed. As these areas grow, retirees are coming to the area desiring a smaller place to enjoy great restaurants and the arts.

One of the most exciting things happening in South Dakota is in the city of Brookings. According to DeVol and Wisecarver, Brookings is a success story. It is home to well-informed entrepreneurs who are highly educated, are establishing tech companies, access to early-stage risk capital, bioscience firms, and advanced manufacturing. Once classified as quasi-rural, it is now classified as a micropolitan area. Current inward migration is creating economic growth. Brookings has created an environment of low taxes, great education, and healthcare. The arts and culture thrive there. This combination attracts young professionals who want to raise children in a safe place (DeVol and Wisecarver, 2018).

A remarkable group of disruptive leaders leads the Oxford-Lafayette County Mississippi Economic Development Foundation. Because of the hard-passionate work of these disruptive leaders, Oxford, Mississippi is touted as the "Cultural Mecca of the South". Writers flock to Oxford to find inspiration in the rich history and the world of William Faulkner. Musicians and artists abound and the culinary experiences do not disappoint. Of course, the presence of the University of Mississippi (Olde Miss) adds to the southern charm of the small town. Oxford is a community rooted in history, where people are as passionate about sports as they are about their favorite authors. Diversity is the norm, and along with southern culture and hospitality make Oxford is unique. Oxford is in the county of Lafayette, in the midst of rolling hills of beauty. The town is filled with boutiques, galleries, and historic landmarks, making Oxford an experience to behold (DeVol and Wisecarver, 2018).

John Maynard, Allen Kurr, and their associates have a philosophy based on the book "The Geography of Jobs" by Enrico Moretti. Moretti lays out how policymakers can learn to lift Americans out of poverty by changing the geography of jobs. By stepping back and looking at the big picture, creativity and innovation will drive change. It is the entrepreneurs that will create wealth, and this is the magical economics of job creation (Moretti, 2012, p. 1). This group of disruptors has taken the plan outlined in Moretti's book and adapted it to their unique situation and has come to some interesting conclusions (Tardif, 2020).

They see themselves in an ocean of tropical fish attracting others to the reef. The real value is to attract entrepreneurs to the reef. Once there, they are given support with money, mentorship, and community backing. By collaborating with the local Arts Council, they host the "Big Bad Business Series", which are workshops providing the tools for success. These workshops resulted in 33 events with over 100 entrepreneurs. In partnership with the University of Mississippi, they are creating the workforce needed for economic development to occur. They have embraced a strategy with five commitments: (Kurr, 2019)

1. Focus on people and attracting people (what do they want)
2. Making sure to support their own people and their ventures
3. Communicate effectively, talk about problems and how to fix them

4. Everything they do is data-driven. Look at the metrics to make decisions
5. Give the "Damn" factor. If you care, it shows in the community.

In May of 2018, The Walton Family Foundation highlighted Oxford as one of the fastest-growing communities in the nation. They gave credit to the Oxford-Lafayette County Economic Development Foundation and the Chamber of Commerce for the "diversified and thoughtful strategic economic development planning", which has led to a 455 percent increase in personal income (DeVol and Wisecarver, 2018).

Mississippi and its state director, Joseph Donavan, who heads up The Entrepreneur Center, focus not only on the arts as part of their creative economy, but on technology as well. Technology is considered part of the innovation and entrepreneurism initiatives. The economic development in the state has grown in the last few years and a paradigm shift has happened in the state. A well-thought-out plan and collaboration effort with small, medium and large businesses are what have made this initiative so successful. In addition, they collaborated with the military on high technology and getting this technology to market (Tardif, 2020).

Ardmore, Oklahoma is an unusual place. It is home to many inventions we use daily. Sylvan N. Goldman owned a grocery store chain and observed that women would stop shopping when their arm carried baskets became heavy. Goldman solved the problem; he invented the first pushable shopping cart. Several other famous inventions and entrepreneurs have come from Ardmore making them one of the most sought-after, cohesively organized micropolitan areas for new businesses to start-up and existing businesses to relocate. In addition, Ardmore recognizes how entrepreneurs and small businesses create a long-term vital economy. The recognition and the culture of hard-working rural labor entice companies to come to Ardmore. The city is the retail location for seven surrounding counties in southern Oklahoma (DeVol and Wisecarver, 2018)

The strategy to diversify the region's economy and increase the knowledge base is attracting new startups. By collaborating with State Universities in fields such as new technology, agricultural biosciences, engineering, and business the new micropolitan areas are full of entrepreneurism that is providing worldwide conferencing and distant educational opportunities. With the geographic locations that provide a quality of life, good schools, low crime and access to major interstate highways, business tax incentives, a stable workforce, economic development, and strategic efforts, the region is attracting and retaining young professionals. These professionals expect a sustainable community so that they can live, work and play.

REFERENCES

Albrecht's, L. (2010). How to Enhance Creativity, Diversity and Sustainability in Spatial Planning: Strategic Planning Revisited. *Urban and Landscape Perspectives*, *9*, 3–25. doi:10.1007/978-90-481-3106-8_1

Cheng, V. M. Y. (2018, April). Views on Creativity, Environmental, Sustainability and their Integrated Development. *Creative Education*, *2018*(9), 719–743. doi:10.4236/ce.2018.95054

Csikszentmihalyi, M. (1996). *Creativity: Flow and the Psychology of Discovery and Invention*. Harper Perennial.

DeVol, R., & Wisecarver, S.J. (2018). *Micropolitan Success Stories from the Heartland*. The Walton Family Foundation.

Maslow, A. (1968). *Creativity in Self-Actualizing People, Toward a Psychology of Being* (2nd ed.). VanNos/Trand Reinhold Company.

Mitchell, I. K., & Walinga, J. (2017). *The creative imperative: The role of creativity, creative problem solving and insight as key drivers for sustainability*. https://www.sciencedirect.com/science/article/pii/S0959652616315037?via%3Dihub

Ramus, C. A. (2001). Organizational Support for Employees: Encouraging Creative Ideas for Environmental Sustainability. California Management Review, 43(3).

Tardif, K.J. (2019). *Developing the Creative Economy through Disruptive Leadership*. IGI Global.

Chapter 14
The World of Quality Control

ABSTRACT

The author's career has gone from plant management, quality control management, natural resource management to corporate affairs and marketing. She had a tea reader tell her to go to the creative side and she would make more money. The tear reader also told her that her marriage would be difficult but would be happy in later years. She should have listened when the tea reader told her to save money. She would need it through her fifties. Why did she not listen? Through autoethnography, this chapter explores the world of quality control.

It would be better if everyone would work together as a system, with the aim for everybody to win. – Dr. W. Edwards Deming

QUALITY CONTROL

The world of quality control has changed drastically from my days when we practiced systems theory and a system of inter-related processes in which people are vital components of the systems. This was called Total Quality Management (TQM). The fourteen points of Dr. Deming are: (Deming Institute, 2021).

DOI: 10.4018/978-1-7998-9691-3.ch014

1. Create a constant purpose for improvement
    a. Plan for the long term
    b. Resist reacting on short-term solutions
    c. Don't just do the same things better- find better things to do
    d. Predict and prepare for challenges if we cannot then be able to act quickly
2. Adopt the new philosophy
    a. Embrace quality through the organization
    b. Put customers' needs first, rather than react to competitive pressure
    c. Be prepared for major changes in the way that business is done
    d. Create your quality vision and values, and implement it
3. Stop depending on inspections
    a. Inspections are costly and unreliable
    b. Build quality into the process from start to finish
    c. Don't just find what you did wrong, eliminate the "wrongs" altogether
    d. Use statistical controls methods
4. Use a single supplier for any one item
    a. Quality relies on consistency
    b. Look at suppliers as partners in quality
    c. Analyze the total cost, not just the initial cost
    d. Use quality statistics to ensure suppliers meet your quality standards
5. Improve constantly and forever
    a. Continuously improve your processes and systems
    b. Life-long learning should be a goal
    c. Use Kaizen or Lean practices to reduce waste and improve productivity and effectiveness
6. Use training on the job
    a. Use on-the-job training for consistency
    b. Build a foundation of common knowledge
    c. Allow employees to understand the "big" picture
    d. Encourage cross-training and encourage teamwork
7. Implement leadership
    a. Expect your supervisors and managers to understand leadership and practice it
    b. Provide support and resources
    c. Understand what each person needs to succeed
    d. Empathize collaboration

   e. Motive and influence everyone to reach their full potential
8. Eliminate fear
   a. Allow people to perform their best
   b. Let everyone know what the goals are
   c. Everyone should feel valued
   d. Leaders should be approachable and listen
   e. Be open and honest
9. Breakdown barriers between departments
   a. Build the concept of the internal customer
   b. Build a shared vision
   c. Develop cross-functional work teams
   d. Focus on collaboration, not consensus
10. Get rid of unclear slogans
   a. Clear and concise message
   b. Do not let slogans replace the effectiveness
11. Eliminate management by objectives
   a. Targets encourage high levels of quality
   b. Measure the process not the people behind the process
12. Remove barriers to pride of workmanship
   a. Let everyone take pride in their work without being compared
   b. Treat everyone the same
13. Implement education and self-improvement
   a. Improve the current skill of workers
   b. Build skills for a more adaptable workforce
14. Make "transformation" everyone's job
   a. Make quality everyone's goal

In addition, we followed the Japanese theory of 7S, this terminology consisted of seven phases: structure, strategy, skills, staff, style, systems, and shared values (Deming Institute, 2021).

This philosophy got us through HACCP (Hazard, analysis, critical control points), SOPs (Standardized Operating Procedures), and tons of documentation. Now Six Sigma, which is a set of techniques and tools for process improvement, are the tools to be lean and efficient in the workplace. Six Sigma has six steps: Define, Measure, Analyze, Improve, and Control. As you go through the program and training you earn certifications. White Belts are for professionals who have not gone through certification training. Below is the rank from the beginning to the end:

- Yellow Belt
- Green Belt
- Black Belt
- Master Black Belt
- Champion

This is thought of as a quality approach that strives for near perfection. It is data-driven to eliminate defects and creates a lean environment. In addition, the top concerns of quality control are to ensure consistency and safety of the product, services, and people. To reduce the variability of materials, boost customer satisfaction, avoid legal liability, reduce waste, and cut costs. When materials are purchased from the same sources, they will not always be the same. Machines are repaired or replaced and might miss maintenance. Food products are the most at risk due to the variability in their processes. Quality control is vital in producing high-quality products and services. When customers are satisfied with the product and services, they stay with them, and new customers will be attracted to them. Furthermore, this lessons legal liability. Quality control issues lead to recalls, and possible lawsuits. Waste is expensive and may occur with a lack of quality control. The best-led companies are those who practice excellence in their quality control.

MEMOIR

This memoir is a series of micro-memoirs over incidents that occurred in my quality control career. I have only mentioned a few, there are too many to write about. I have said many times in this book, I have met the best and worst of people in my profession. Unfortunately, it seems like the worst has been involved in quality control. I do not mean dark people, but people who thought they could make an easy buck.

The "lipstick case" was an interesting quality control complaint. A woman from New York City called to say she purchased a bottle of water from a cart in Central Park, and it had red stuff on the rim of the bottle. Thank God, she noticed it before she drank it. I ask if I could have a delivery guy pick up the bottle and give her a case of water for her trouble. She was more than happy to do so. After I received the bottle, I sent it to a laboratory to be tested. It came back and twenty-four-hour lipstick. Not the type of lipstick that would wash off easily.

I had used an off-duty New York police detective before, and he was great to work with. So, I called and asked him to investigate this. At that time, it was hard to tell. I did not think the woman seemed like the type to do something like this herself. After two weeks of investigating, he informed me he tracked it down to a cart vendor in Central Park, who was taking the bottles home, filling them with New York City tap water, and using black market caps to seal them. Of course, without commercial cleaners, he could not remove 24-hour lipstick. He was arrested for tampering with a food product, which is a felony. He could have made someone very sick. This was a serious issue.

The "fish tank story" was another case, that was easier to solve than the lipstick issue. I had a 1.5-liter bottle of water sent to me that was half full. The customer said she purchased it this way. It looked like fish tank water was inside of it and stunk like dirty fish tank water. She wanted thirty-five thousand dollars to recompense her for getting sick after drinking the water. No person sane person would drink this water. By this time, I was a seasoned quality control manager. I called the customer and explained the production date on the bottle, and the quality logs that went with the production date. I informed her that I had a supply of water from all production times in inventory, and they were all pure water. This had to be tampering. I informed her I would be calling the FBI, to come to take the product, and they would be calling on her, and investigating the tampering of a food product which is a felony and by the way, so is extortion! I asked her if she wanted to recant her story, which she did, and slammed the phone down.

Another funny story was taking samples from a spring. One afternoon I was watching bottles at the start of a production line and noticed they had a lavender hue. I did a double take! We stopped production and started pulling bottles off the line and putting all finished products on a pallet. We cleaned the equipment and changed springs and products. Water is all different, it is made up of various changes with dissolved solids, minerals, ph., and much more. Different springs were used for different labels. People who purchase products seem to like different springs. Most customers liked a dissolved solid in the water around 35 to 70 ppm. This particular spring was always around 65ppm. We tested the water and found the magnesium was slightly elevated. Not out of compliance, but not what I wanted in the product.

I decided to go up to the spring early in the morning and take a look and draw some samples. This particular spring had no borehole, only a spring catchment. The catchment was in a spring house, that was spotless and protected. I got there early, and everything looked great, and I took my samples. The spring was at the top of a mountain. The area had high rainfall levels in the previous week, and this probably contributed to the slightly elevate magnesium. When water filled the bottles the action slightly aerated the water and a slight lavender hue was noted in the water.

Coming down from the mountain was a beautiful walk, and at the bottom were blackberry bushes full of big thumb-size blackberries. I decided to bring some back to the plant and put them in the break room. I went down to where our tanker trucks loaded and talked to one of the drivers, and we rinsed out a five-gallon bucket to use to pick the blackberries. I was engrossed in picking blackberries and the tanker driver was just watching, enjoying the nice morning. I stuck my hand through a bush reaching for a big blackberry when I touched something wet. I pulled the branches apart, and there in front of me was a black bear. I screamed, and ran off, the bear looked as scared as me and ran up the mountain. The tanker driver retrieved my bucket. Yes, I was nervous to go back up, and later felt awful, because I realized in was just a junior bear. Junior bears tend to get in trouble because they usually miss their mothers when it's their first year alone. They seek out company, so he just wanted to enjoy the blackberries with me. By day's end, the employees enjoyed the blackberries. The water was fine, and in a few days the magnesium went down, and no lavender water was seen again.

The hardest issue I had was trying to get a new development permitted in an adjoining state. It was winter, and the regulator was being difficult and demanded water testing to be done the next day. I informed her that we were having extreme weather, but she would not relent. I think she thought there was no way to pull it off, and in all honesty did not want to permit us. Some people just like power over other people. When I woke up in the morning it was -38 degrees F. I mean temperature, not windchill. The tractors that were plugged in would not start. By the time I got to the plant, it had not changed. I had the bottles, a large box for each spring. There were four boreholes to test. I dressed in layers of clothes with an insulated coverall. A vendor, a super nice guy volunteered to meet me there and take the bottles to and from the boreholes and the tanker loadout station. This way they would not freeze. We would keep them there, where it was warm until I was done. Then I would take them to Fed Ex. I would be on snowshoes, walking to each borehole.

It was the coldest day of my life. I still do not know how I did not get frostbite. We started at 7:30 in the morning and ended at 5:00 in the afternoon. All day walking on 3 feet of snow with snowshoes. I was totally exhausted. It was getting dark, and I only had two boxes done. They were not labeled. Since I was out in the woods, I decided it best to go finish at Fed Ex. I pulled into the Fed Ex office and as I was bringing the bottles in, I looked at my reflection. It was not pretty. I told the Fed Ex employee I had been outside since 7:30 and I asked if I could finish labeling and taping in their office. She was amazing and one of the kindest people I have ever met. She said for me to sit and take off some of my layers. She got me cup of hot coffee and said she would do the work. All had to do was tell her what to do. She saved me that evening. I sent her flowers the next day and wrote a letter to her corporate office. I hope they rewarded her.

I took great pleasure faxing the regulator my Fed Ex receipt. I reminded her by phone, that I was a single parent at the time with four children and got home at 8:30 from being in the deep woods, with deep snow on the coldest day in years. She asked me why I did not have one of my quality control technicians do it. I told her I would not have put anyone, let alone one of my technicians, through that. The toughest jobs are left to the leaders. I wish I could have expressed that to some men above me. The plant manager had been calling my house every thirty minutes from six o'clock on and was very relieved when I answered the phone.

When you are willing to be in the thick of it, you earn tremendous respect. Early on at that job, I learned this. It was another January cold day, for the previously mentioned regulator I had to take a sample from a spring-fed pond. One of our maintenance people and I went out to the pond where we knew spring vents were underneath. You might be thinking there should be no ice over the moving water, right? Think again. This ice is at least 18 to 24 inches thick; they can drive tractor trailers on this ice. One pond has a railroad track laid down in January and February to drive a train across the ice, which is a shortcut, and better than driving on two-lane curvy roads. These ponds are called kettle ponds, because they are like a teapot on the stove, with the springs bubbling up underneath them.

The ice auger went down below the ice, and a water sample had to be taken there. The maintenance guy looked at me disgusted, and said," I guess you want me to go into the hole". I said no, I will go, but because I am short, you will have to hold my ankles. Please do not let go, because I have children at home. He held my ankles, and I got the sample. I had brought a towel and lap blanket to wrap up in to go back to the plant where I would change clothes. He was impressed and told the entire plant about it. Later in the day, walking through the plant someone commented on it. I said, do not underestimate a farm girl from Kansas, and by the way we do not take prisoner's either. That went through the plant also and did wonders for my reputation. Upon my retirement, I was no longer in the plant having moved on to the corporate office, the plant employees sent me the most beautiful flowers. They also gave me the first gift they ever gave a manager. For Christmas that year they gave me amethyst earrings. Normally I could not take them, but my boss let me so I would not hurt their feelings. Many memories, from wonderful lovely people.

As a single mom my children often did their homework in my office or the breakroom. They were always treated with respect, and I was very proud of them, they were well behaved. So, when doing something extra, a fun project where I thought they could learn, I included them. I was visiting with a state legislator, and she was telling me the loons were leaving the lake. I asked if she knew why, and she said the lake level was fluctuating. The loons would build their nest in the spring, but in summer the water level would drop, and the loons could not reach their nest. Loons have wide feet with short legs. When they take flight, it takes them a while to get going but walking up to their nest is hard for them, so they need their nest close to the water. I called the state and asked them what we could do to help our friends the Loons. They told me we had to build loon nest.

This was not hard, but something I could not do alone. My son and I purchased the materials, and spent a Saturday, on a March day building loon nests. You need wood that can handle water, chicken wire, and a shovel. We made a raft from the wood, stapled the chicken wire to the wood and put an anchor on the raft. Then on top of the chicken wire we put in mud, dirt, and native plants. When done it looked like a miniature island? We put it far enough out that it would float with the lake level. We returned home and cleaned up and started monitoring our project with binoculars. I phoned the state legislator and told her what we had done, since she lived on the edge of the pond, and she started monitoring with her binoculars. It worked, we had seven pairs of loons come back to the lake that summer.

This state legislator had a bread and breakfast on this lake. Our corporate people stayed their often. Onetime our CEO and an Executive Vice President were staying at her bed and breakfast observing the loons. She told them they would not be there if not for my son and me. Since I had raised the quality score in the plant to the best in the company, when before I came it was the worst score in the company, they honored me with the President's Award that year. This award was for job performance and community service.

This story however does not end with the award. There was still the issue of lake levels. When we were building our Loon nest, we had a resident, who owns property on the lake tell us he felt it was our fault the lake levels were going up and down, and fish were not as plentiful. I had to address this. After a few weeks of looking into the situation and taking water samples at his end of the lake, which was the opposite side of ours, I found the town had their salt mound on the side of the lake, and every time it rained the salt was leaching into the lake. In addition, at the end of the lake close to where he lived was an abandoned hydro dam. To mitigate the problem, I received bids to move the salt mound and repair the dam. The dam was sixty-nine thousand dollars to repair and to move the mound was another ten thousand dollars. I asked the town if we could do this, and they approved the projects. We did it, and the lake level stabilized, and fish gradually rebounded.

I did this, to save time and money. If the public thought we were at fault, they would hire lawyers and I would spend a million dollars to prove this was not my issue. The lawyers would make all the money. This was a better way to solve the issue. We gave to the town and our neighbors. I was tired of hearing our lawyers brag of how much money they were making from us.

A few years later we had the ice storm of the century. Everyone was worried about the apocalypse coming. I knew this was a hundred-year weather event. A brochure from 1898 showed pictures of the ice storm of the century. People were in their long skirts, and furs, standing next to downed trees covered in ice. The Inn, that was part of a great resort that included an historic spring was a popular destination. The historic spring house and bottling plant were now museums. Back in the day modern transportation did not exist, but I am sure their life was stopped, just like ours was from the ice storm. Grocery stores were on a cash only basis because they had no power. Power was out for days, weeks and in some cases months in Canada.

We brought in generators to run the plant, so we could send tankers of water to the area hospitals, dialysis centers, and for residents we took truckloads of water to area fire stations that people could come and get for home use. In addition, I had our plow trucks clear roads because the town was not plowing, and the electrical company could not get in to restore power. We did this to be a good neighbor, but we knew the importance of good clean water for health. There is always some criticism of our industry, but this is an example of how our water was needed in time of devastation. Even in normal times many people do not have access to clean, healthy, good tasting water.

Water is the lifeblood of our planet. Alternative sources of water are needed, at least until everyone on the planet has access to clean, healthy, pure water. *"It is life, I think, to watch the water. A man can learn so many things."* ***Nicholas Sparks***

REFERENCES

Campus, G. (2021). *Quality Control.* Grey Campus. https://www.greycampus.com/blog/quality-management/quality-control-and-quality-assurance

Deming Institute. (2021). *Harnessing the power of genuine collaboration.* Deming Institute. https://deming.org/

Sigma, S. (2021). *Helping People Find Solutions.* Six Sigma. https://www.6sigma.us/

Chapter 15 Glossary: The Science of Water

ABSTRACT

This glossary is to be a guide for those who are laypeople or students who are learning water and sustainability. For educators and researchers, the author's hope is that this provides a tool to teach from or a quick reference on the definitions of water management and sustainability. We are the water planet, and the author's hope is that as a society we become more knowledgeable about water and sustainability. Recently scientists have discovered the Earth has a pulse. Rampino and Zhu havesuggested that global geologic events are generally correlated and seem to come in pulses with an underlying 27.5-million-year cycle. These cycles may be related to the dynamics of tectonics and mantle plumes or might alternatively be paced by astronomical cycles associated with the Earth's motions in the Solar System and the Galaxy. We are still learning about our beautiful planet and its place in our galaxy.

WHAT IS GROUNDWATER?

The simplest definition is that groundwater is water contained in saturated soil and rock materials below the surface of the earth. Groundwater is recycled water that is related to all the other water on earth by the process called the hydrologic cycle.

The source of groundwater is precipitation. When rain falls on the surface of the ground, some of it runs off the land into lakes and streams (runoff),

DOI: 10.4018/978-1-7998-9691-3.ch015

Figure 1. Water cycle
(UAGS, 2022).

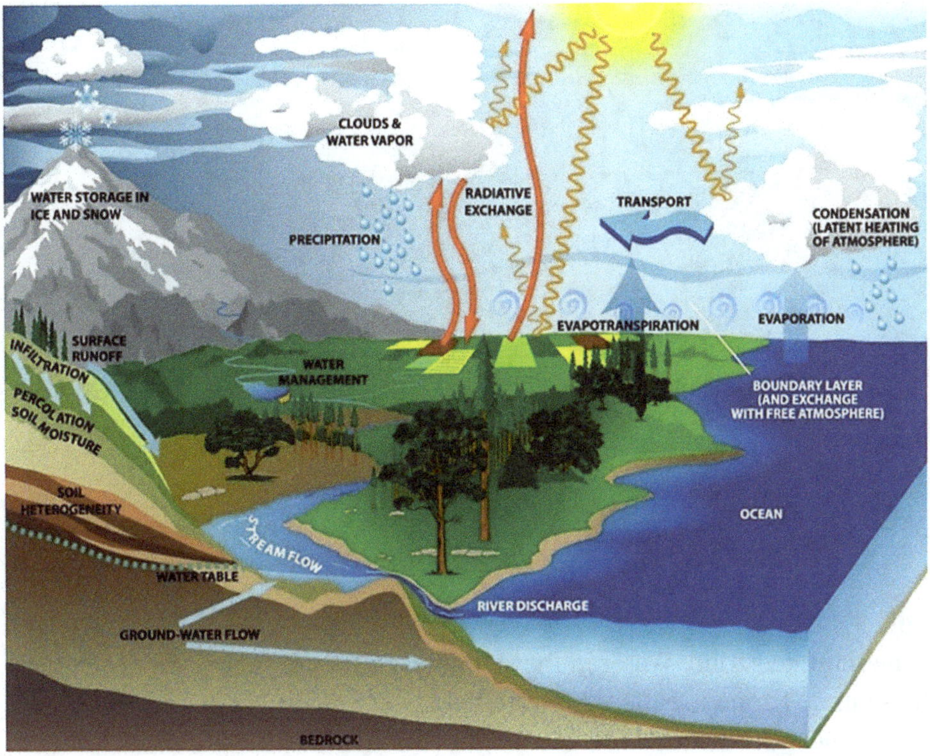

and some soak into the ground (infiltration). The water soaking into the ground first goes through the unsaturated zone, which is the layer of sand and gravel. In the unsaturated zone, some of the spaces between the soil particles are filled with water, and some are filled with air. Some of the water in the unsaturated zone will be lost to the atmosphere by evaporation.

Soils in the unsaturated zone are able to hold water in small pores against the force of gravity because of surface tension or cohesion, which is the attraction that water molecules have for one another, and because of the attractive forces between the soil particles and the water molecules. Water in larger pores is subject to the force of gravity and is the source of water that moves downward to become groundwater.

Below the unsaturated zone, the water reaches a zone in the sand and gravel where all the cracks and spaces in the soil or rock are filled with water. This

Figure 2. Groundwater
(USGS, 2022).

is the saturated zone. Water in the saturated zone is groundwater. The top of the saturated zone is called the water table.

Groundwater moves slowly downgradient between the grains of soil or in cracks in rock until it reaches a point where it can discharge at the surface, such as a lake, stream or wetland, or until it is withdrawn from a well. It then becomes surface water again. The groundwater that we use today may have traveled through the hydrologic cycle hundreds or thousands of times since the earth was formed.

HOW IS WATER STORED UNDERGROUND?

Groundwater is stored underground in the pore spaces of saturated soil and rock materials. An underground unit of soil or rock, which can yield a significant quantity of groundwater to wells, is called an aquifer. Three of the most common aquifers are: a sand and gravel aquifer, limestone aquifer, and a sandstone aquifer. Groundwater flows through interconnected pore spaces in aquifers; groundwater may flow at different rates either horizontally or vertically, because of differences n composition or in properties.

Figure 3. Aquifer
(USGS, 2022).

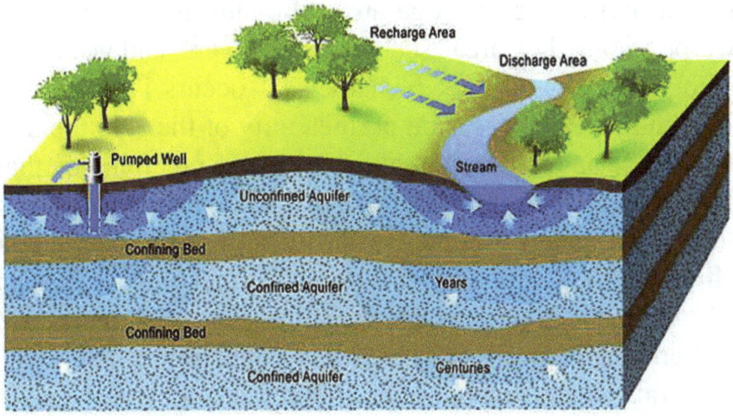

Aquifers may be separated by layers that do not transmit much water. These layers are called confining layers (aquitards). If a confining layer exists above an aquifer that is fully saturated, this aquifer is then a confined or artesian aquifer. Aquifers with a confining layer above them are called unconfined aquifers or water table aquifers.

When a well is drilled to penetrate any aquifer, water will enter the well casing. In an unconfined aquifer, the water level will stabilize in the well at the top of the saturated zone, which is called the water table. In a confined aquifer, water in the well will rise above the top of the aquifer. The level to which the water rises is called the potentiometric surface. The potentiometric surface of the confined aquifer is analogous to the water table of the unconfined aquifer. If the potentiometric surface is above the surface of the ground, a flowing well or spring may result.

WHERE DOES GROUNDWATER COME FROM AND WHERE IS IT GOING?

Once water percolates down to the water table to become groundwater, it starts to move slowly downgradient toward an area where it can discharge at the surface of the ground. Those areas where water enters the groundwater system are called recharge areas, and those areas where groundwater reappears at the surface of the ground are called discharge areas.

Groundwater recharge areas are usually located in upland areas. However, in areas where topographic differences are small, most of the land area can be considered part of the recharge area. Groundwater discharge areas are normally typographic low areas such as lakes, rivers, and wetlands.

The amount of groundwater recharge that occurs is related to several factors, including the porosity and permeability of the soil, the topography of the land surface and the amount of timing of the precipitation that occurs.

Porosity is the ratio of the volume of the void space in a material to the total volume of the material. Porosity is influenced by the size and the shapes of the individual grain size withing the soil. Permeability is the capacity of a material to transmit fluids, which is mainly controlled by the size of the pore spaces and the degree to which they are interconnected.

Timing of rainfall is also important. If rain fall at the time when crops and vegetation are actively growing and using water, very little may make its way to the saturated zone. In fact, in many areas the major recharge periods occur in spring and fall, when precipitation is greater, and vegetation or crops are not actively intercepting and using as much water.

Topography influences the rate of groundwater recharge as well. In steep terrains, more water may run off the land into surface water than in flatter terrains.

DARCY'S LAW ON THE FLOW OF WATER THROUGH POROUS MATERIALS

In the 1850's, a French engineer, Henry Darcy, studied aspects of the movement of water through porous materials such as those found in groundwater aquifers. His results have become widely used in groundwater study and are known as Darcy's Law. Darcy found that the amount of water that flowed through porous materials in a cylinder was directly related to the correctional area of the cylinder, and inversely related to the length of the cylinder. He also found that the flow was directly proportional to the hydraulic conductivity of the material.

Darcy's Law tells us that the rate of groundwater flow is predictable when we know something about the aquifer that groundwater is flowing through. The laws of physics tell us that for this groundwater flow to occur, energy is required. Groundwater flows in response to differences n energy levels. The laws of physics tell us that all substances have energy in mechanical, thermal,

Figure 4. Water
(Dream team, 2022).

and chemical forms. Nature is constantly trying to equalize energy levels, and so groundwater moves from areas of high energy to areas of lower energy. Since most groundwater has negligible thermal or chemical differences, mechanical energy is the most significant energy source for groundwater flow.

The total mechanical energy that makes groundwater flow is called hydraulic head. The hydraulic head is defined as the sum of three factors: kinetic energy (the energy of motion), gravitational energy, and pressure energy. The gravitational energy comes from the difference in elevation between the recharge area and the discharge area. Therefore, it is also called the elevation head. The pressure energy is also called the pressure head. Hydraulic head is expressed in terms of energy per unit weight of water. The force potential (energy per unit mass) is multiplied by a mass of water, we get the physical expression of energy (the joule). The force potential is the driving energy behind groundwater flow.

Elevation head and pressure head are both important terms in this equation, because they both contribute to the hydraulic head, but the relative proportion of each is unimportant. Water can flow "uphill" if the pressure head is greater than the elevation head.

WHAT PROPERTIES OF WATER MAKE IT UNIQUE?

Water is a vital resource for all living things. It is believed that life originated in water. The bodies of living organisms are mainly composed of water. All living things need water to survive. The many unique properties of water cause it to have a tremendous impact on our physical environment as well.

Water can dissolve more substances in greater quantities than any other liquid.

Water has high specific heat, which is the number of calories required to raise the temperature of water. This helps us to maintain the temperature of own bodies despite extremes of cold and heat. This same high heat capacity allows large lakes and oceans to moderate the temperature of nearby land.

Water has the highest heat of evaporation of any liquid. This means that the large amounts of heat energy are needed to evaporate even small amount of water. The release of this energy is the driving force behind the world's weather systems. In addition, this property helps to keep the body temperature of living organism's constant, since a large amount of heat can be dissipated by vaporizing a small amount of water.

Water also has a high heat of fusion, meaning that a large amount of heat energy is given off as water changes form from liquid water to solid ice. This property also moderates the temperature of bodies of water in the wintertime.

Water reaches its maximum density at 39 degrees F. Therefore, ice is less dense than water and floats in water. If this were not true, lakes would freeze over from the bottom up, making them very difficult to thaw in the spring and drastically changing the habitat of aquatic organisms. Water molecules are dipolar: that is the positively charged hydrogen ions are on one side of the molecule and the negatively charge oxygen ion is on the other side. The polarization leads water molecules to be attached to each other by a phenomenon called hydrogen bonding. Hydrogen bonding causes water molecules in the atmosphere to coalesce and fall as raindrops.

When water is compared to compounds of similar molecular structure, it has unusually high boiling and melting points, attributable to hydrogen bonding. Hydrogen bonding also causes water molecules to be strongly attracted to mineral surfaces. Water easily wets and penetrates small rock fractures. Freezing and expansion of the water in cracks causes much of the weathering that produces our soils. The polar nature of water molecules is largely responsible for its ability to dissolve many other substances. Water is often called a universal solvent for this reason.

GLOSSARY OF HYDROGEOLOGIC TERMS

A

Abandonment: Generally refers to the surrender, relinquishment, or cession of property or of right. Applied to water rights, defined as an intentional relinquishment of a known right to use water by virtue of nonuse.

Acequia: A Spanish work used the Southwest, USA to mean a community irrigation ditch or canal.

Acid: A substance that has a pH f less than 7, which is neutral. Specifically, an acid has more free hydrogen ions (H+) than hydroxyl ions (OH-).

Acre-Foot: The quantity of water required to cover one acre of land to a depth of one foot, which is the equivalent to 325, 851 gallons of water.

Adhesion: The attraction of water molecule to solid surfaces.

Adjudication: A formal judicial proceeding to determine the extent and validity of a water right.

Alkaline: Sometimes water or soils contain an amount of alkali (strong basic) substances sufficient to raise the pH value above 7.0 and be harmful to the growth of crops.

Alkalinity: The capacity of water for neutralization an acid solution.

Alluvium: Deposits of clay, silt, sand, gravel or other particulate material that has been deposited by a stream or other body or running water in a streambed, on a flood plain, on a delta, or at the base of a mountain.

Anion: A negatively charged ion that migrates to an anode.

Appropriation: A grant of authority from a state entity to divert, store, or use the public waters of the state. The word appropriation is often used interchangeably with the terms right and permit.

Appropriation Doctrine: The system western water law pertaining to the first-in-state right of taking or diverting water from a watercourse and applying it to a beneficial use.

Appropriator: A person who take water from a watercourse and applies it to a beneficial use.

Aquaculture: Farming of plants and animals that live in water, such as fish, shellfish, and algae.

Aqueduct: A pipe, conduit or channel designed to transport water from a remote source, usually by gravity.

Aquifer: A saturated underground geologic formation of rock or other porous material capable of storing water and transmitting it to wells or springs.

Aquitard: A zone with the Earth that restricts the flow of groundwater from one aquifer to another. Comprise of layers of clay or non-porous rock with low hydraulic conductivity.

Area of Origin Protection: Refers to law, regulation, or policy that provides some measure of protection for a region from which an interbasin water transfer is made.

Artesian Aquifer: See confined aquifer.

Artificial Recharge: A process where water is put back into groundwater storage from surface water supplies such as irrigation, or induced infiltration form streams or wells.

B

Base: A substance that has a pH of more than 7, which is neutral. A base has less free hydrogen ions (H+) than hydroxyl ions (OH-).

Base Flow: Sustained or fair-weather flow of a stream, whether affected by the works of man. The discharges entering stream channels from ground water of other sources.

Base of Origin: The river basin in which surface water naturally occur.

Bedrock: The solid rock beneath the soil and superficial rock. A general term for solid rock that lies beneath soil, loose sediments, or other unconsolidated material.

Beneficial Use: The use of water for the benefit of the appropriator, other persons, or the public, defined more specifically within each state's water law. It may include (but is not limited to) water used for agriculture (including stock water), domestic, fish and wildlife, industrial, mining, municipal, power, and recreational use.

Borehole: A circular hole made by boring, especially a deep hole of small diameter.

C

Call of River: To make a request or demand that water right holders on a watercourse appropriate water only in accordance with the priority date ranking.

Capillarity: The phenomenon by which water rises in small pores into the unsaturated zone above the water table, because of adhesive and cohesive forces.

Cation: An ion having a positive charge and, in electrolytes, characteristically moving toward a negative electrode.

Certified Water Right: An official, written assurance that the necessary requirements to perfect or finalize a water right have been completed (e.g., diversion and beneficial use).

Cohesion: The attraction of water molecules to each other because of hydrogen bonding. Also know as surface tension.

Commercial Water Use: Water used for motels. Hotels, restaurants, office buildings, other commercial facilities, and institutions. Water for commercial uses come both public supplied sources, such as a country water department, and self-supplied sources, such as wells.

Condemnation: The process of taking private property for public use through the power of eminent domain. The fifth Amendment to the U.S. Constitution requires that just compensation be paid.

Condensation: The process whereby water is changed from gas to liquid.

Conditional Permit: A water use permit that is dependent upon or granted subject to a condition. For example, a temporary permit may authorize water use for a specific quantity of water in a specified place for a specified time and under specified conditions.

Cone of Depression: The zone around a well in an unconfined aquifer that is normally saturated but becomes unsaturated when the well is pumped.

Confined Aquifer: An aquifer that is bonded above and below by confining layers which transmit water significantly slower that than of the aquifer. The water level in a well tapping a confined aquifer will rise above the top of the aquifer because the confined aquifer is under pressure. Also called artesian well.

Confining Layer: A layer of lower permeability material that overlies an aquifer. Sometimes called an aquitard.

Conjunctive Management: Integrated management and use of two or more water resources, such as an aquifer and a surface water body.

Conjunctive Use: The integrated use and management of hydrologically connected groundwater and surface water.

Conservation: Water saving methods that serve to increase water supply by decreasing demands for water.

Consumptive Use: Water withdrawn from a supply that, because of absorption, transportation, evaporation, or incorporation n a product, is not returned directly to a surface water or groundwater supply; hence, water that is lost for immediate further use.

Conveyance Loss: Water that is lost in transit from a pipe, canal, or ditch by leakage or evaporation. Generally, the water is not available for further use; however, leakage from an irrigation ditch, for example, may percolate to a groundwater source and be available for further use.

Cubic Feet per Second (cfs): A standard unit of measure for flowing water. One cfs is equal to 448.8 gallons per minute.

D

Degree: The judgement of a court, an official order, or settlement.

De Minimis: Derived from the Latin phrase De minimis non curat lex, meaning that the law does not care for, or take notice of, very small or trifling matters. De minimis water uses are those considered by law too insignificant to notice.

Depletion: The withdrawal of water from surface or groundwater reservoirs at a rate greater than the rate of replenishment.

Desalination: The removal of salts from saline water to provide freshwater. This method is becoming a more popular way of providing freshwater to populations.

Dilution: A greater volume of groundwater being mixed with the same amount of contaminant.

Discharge: The removal of groundwater from an aquifer or lakes, streams, or wetlands, by pumping a well, or by leakage to another.

Dispersion: Water carrying a contaminant becoming more separated by following a variety of distinct flow paths through an aquifer. Dispersion has two components: mechanical mixing and diffusion.

Dissolved Minerals: Mineral (salts included) that are in solution in water.

Dissolved Oxygen: The amount of oxygen in solution in water.

Ditch Rider: Popular name for a water commissioner or local person who is responsible for overseeing water allocation among all water users and for enforcing the priority system on a give watercourse.

Diversion: The removal of water from a natural watercourse by canal, pipe, well, or other conduit for transfer to another watercourse or for application on the land; also called withdrawal.

Domestic Water Use: Water used for household purposes, such as drinking, food preparation, bathing, washing clothes, dishes, dogs, flushing toilets, and watering lawns and gardens. About 85% of domestic water is delivered to homes by a public water supply facility, such as a county water department. About 15% of the Nation's population supply their own water, mainly from wells.

Drainage Basin: The land area from which water drains into a river, also called a watershed or river basin.

Draw Down: (groundwater) The lowering of the water level in a borehole as a result of pumping; reduction of the pressure head as a result of the pumping; reduction of the pressure head as a result of the withdrawal of a water from a borehole; (hydraulic) In a stream or conduit, the difference between the water surface elevation at a constriction; (water) The distance by which the level of a reservoir or lake is lowered by the withdrawal of water.

Drip Irrigation: A common irrigation method where pipes or tubes filled with water slowly drip onto crops. Drip irrigation is a low-pressure method of irrigation and less water is lost to evaporation than high-pressure spray irrigation.

Duty of Water: The amount of water necessary and economically used, to successfully grow crops without unnecessary water loss or waste.

E

Effluent: Water that flows from a sewerage plant after it has been treated.

Elevation Head: The difference between the recharge area and the discharge area.

Eminent Domain: The right of government to acquire private property for public use, even from an unwilling owner, upon payment of just compensation to the owner.

Enforce: To put something as a law into effect. Water rights enforcement generally refers to allocating water resources in strict accordance with priority dates associated with each water user's water right, in accordance with terms and conditions specified.

Environmental Protection Agency: An independent agency of the federal government formed in 1970 and responsible for pollution abatement and control programs, including programs in air and water pollution control, water supply and radiation protection, solid and toxic waste management, pesticides control and noise abatement. Federal agency that permits coordinated and effective governmental action to assure the protection of the environment by abating and controlling pollution on a systematic basis.

Erosion: The process in which a material is worn away by a stream or liquid (water) or air, often due to the presence of abrasive particles in the stream.

Esker: A long, narrow ridge of coarse gravel deposited by a stream flowing in an ice-walled valley or tunnel in a decaying glacial sheet.

Estuary: A place where fresh water and saltwater mix, such as a bay, salt marsh, or where a river enters an ocean.

Eutrophication: The process by which results in a water body which is very rich in nutrients, seasonally deficient in oxygen, and generally shallow.

Evaporation: The process whereby water is changed from a liquid to a gaseous state.

Evapotranspiration: Water dissipated to the atmosphere by evaporation from moist soil and plant transpiration.

F

Flood: An overflow of water onto lands tat are used or usable by man and not normally covered by water. Floods have two essential characteristics. The inundation of land is temporary; and the land is adjacent to an inundated by overflow from a river, stream, lake, or ocean.

Flood, 100 year: A 100-year flood does not refer to a flood that occurs once every 100 years, but to a flood level with a 1 percent chance of being equaled or exceeded in any given year.

Floodplain: The land bordering a river that is subject to flooding during any given period.

Flood Stage: The elevation at which overflow of the natural banks of a stream or body of water begins in the reach or area in which the elevation is measured.

Flowing Well/Spring: A well or spring that taps groundwater under pressure so that water rises without pumping. If the water rises above the surface, it is known as a flowing well.

Flow Rate: The speed or rate at which water is taken from a watercourse or the speed at which it flows past a point (e.g. gallons per hour of minute, cubic feet per second).

Forfeiture: The loss of a water right due to nonuse for a specified time; varies by state. Can occur voluntarily.

Freshwater: Water that contains less than 1,000 milligrams per liter (mg/L) of dissolved solids; generally, more than 500mg/L of dissolved solids is undesirable for drinking and many industrial uses.

G

Gage Height: The height of the water surface above the gage datum (zero point). Gage height is often used interchangeably with the more general term, stage, although gage height is more appropriate when used with a gage reading.

Gaging Station: A site on a stream, reservoir, or other body of water where observations and hydrology data are obtained. The U.S. Geological Survey measures stream discharge at gaging stations.

General Stream Adjudication: Involves a judicial proceeding to determine the extent and validity of all water rights within a given geographic area (e.g., one or more river basins or statewide.)

Geophysics: The science of the earth with respect to its structure, composition, and development.

Geyser: A geothermal feature of the earth where there is an opening in the surface that contains superheated water that periodically erupts in a shower of water and steam.

Giardiasis: A disease that results from an infection by the protozoan parasite Giardia Intestinalis, caused by drinking water that is either not filtered or not chlorinated. The disorder is more prevalent in children that adults and is characterized by abdominal discomfort, nausea and alternating constipation and diarrhea.

Glacier: A huge mass of ice formed on land by the compaction and recrystallization of snow, that moves very slowly downslope or outward due to its own weight.

GPM/Ftdd: Gallons per minute per foot of drawdown. A unit to express the specific capacity or productivity of a borehole.

Greywater: Wastewater from clothes washing machines, showers, bathtubs, handwashing, lavatories, and sinks.

Groundwater: Water that occurs beneath the land surface and fills the pore spaces of the rock material in which it occurs, also called percolating water.

Groundwater Mining: The withdrawal of water from an aquifer at rates more than net recharge or replenishment. Eventually, the underground supply will be exhausted or will not be economically feasible to pump.

Groundwater Recharge: Inflow of water to a groundwater reservoir from the surface. Infiltration of precipitation and its movement to the water table is one form of natural recharge. Also, the volume of water added by the process.

Grout Interval: That portion of the section filled with cement between the wall of the borehole and casing.

H

Hardness: A water quality indication of the concentration of alkaline salts in water, mainly calcium and magnesium. If the water you use is "hard" then more soap, detergent or shampoo is necessary to raise a lather.

Headwater(s): (1) The source of upper reaches of a stream; also, the upper reaches of a reservoir. (2) The water upstream from a structure or point on a stream. (3) The small streams that come together to form a river. Also maybe thought of as all parts of a river basin except the mainstream river and main tributaries.

Heat of Fusion: The energy gives off as water changes forms from liquid to solid.

Heat of Vaporization: The energy consumed as water changes from a liquid to a gas.

Hydraulic Conductivity: "permeability coefficient"! The rate of flow of water in gallons per day through a cross section of one square foot under a unit of hydraulic gradient, at prevailing temperature (field permeability coefficient) or adjusted for a temperature of 60 degrees F.

Hydraulic Head: The height of the free surface of a body of water above a given subsurface point. The water level at a point upstream from a given point downstream.

Hydraulic Impact: The change in water level or pressure head caused by withdrawal of water.

Hydroelectric Power Water Use: The use of water in the generation of electricity at plants where the turbine generators are driven by falling water.

Hydrogen Bonding: The cohesive force caused by the polar nature of water, in which the oxygen atom in one water molecule is attracted to the hydrogen atom of another water molecule. This bonding causes water to have many of its unique properties.

Hydrogeologic Cycle: The circulation of water from the sea, through the atmosphere, to the land and then back to the sea by overland and subterranean routes or directly back into the atmosphere by evaporation and transpiration.

Hydrology: The science of the behavior of water in the atmosphere, on the earth's surface, and underground.

I

Impervious Rock: Layer of rock generally considered under normal conditions to be incapable of being penetrated.

Indicator: A chemical or biological parameter that can be used to indicate the possible presence of other contaminants.

Industrial Water Use: Water used for industrial purposes in industries such a steel, chemical, paper, and petroleum refining. Nationally, water for industrial use comes from mainly 80% self-supplied sources, such as local wells or withdrawal points in a river, but some water comes from public supplied sources, such as the country/city water department.

Infiltration: Flow of water from the land surface into the subsurface.

Injection Well: Refers to a well-constructed for the purpose of injecting treated wastewater directly into the ground. Wastewater is generally (pumped) into the well for dispersal or storage into a designated aquifer. Injection wells are generally drilled into aquifers that do not deliver drinking water, unused aquifers, or below freshwater levels.

Instream Flow: The amount of water remaining in a stream, without diversion, that is required to maintain a particular aquatic environment or water use.

Instream Use: Use of water that does not require withdrawal or diversion from its natural watercourse, for example, the use of water or navigation, recreation, hydroelectric production, and support of fish and wildlife habitat. A non-consumptive use of water.

Interbasin Transfer: The physical transfer of water from one watershed to another.

International Water Treaty: An agreement between the United States and a foreign nation state dealing with a water resource involving both nations. Such treaties supersede state law. The United States has treaties with Mexico and Canada relating to shared water resources.

Interstate Water Compact: An agreement between two or more states dealing with competing demands for a water resource beyond the legal authority of one state alone to solve. Such agreements require the consent of Congress.

Irrigation: The controlled application of water for agricultural purposes through manmade systems to supply water requirements not satisfied by rainfall.

J

Just Compensation: Compensation that is fair to both the owner and the public when property is taken for public use through condemnation or eminent domain. Criteria applied may include the market value of the property and the resulting damage to the remaining property of the owner.

K

Kilogram: One thousand grams.
Kilowatt Hour: A power demand of 1,000 watts for one hour. Power company utility rates are typically expressed in cents per kilowatt-hour.

L

Leaching: The process by which soluble material in the soil, such as salts, nutrients, pesticide chemicals or contaminants are washed into a lower layer of soil or are dissolved and carried away by water.
Lens: A geologic deposit bounded by converging surfaces (at least one of which is curved), thick in the middle and thinning out towards the edges, resembling a convex lens.
Lenticular Zone: A lens shaped zone in the subsurface with limited lateral extent.
Lentic Waters: Ponds or lakes (standing water).
Levee: A natural or manmade earthen barrier along the edge of a stream, lake, or river. Land alongside rivers can be protected from flooding by levees.
Lithologic: pertaining to the description and classification of rock, clay, gravel, sand, etc.
Livestock Water Use: Water used for livestock watering, feed lots, dairy operations, fish farming, and other on-farm needs.
Lotic Waters: Flowing waters, such as streams and rivers.

M

Mainstream: The principal watercourse of a river, excluding any tributaries.
Maximum Contaminant Level (MCL): The designation given by the U.S. Environmental Protection Agency (EPA) to water-quality standards promulgated under the Safe Drinking Water Act. The MCL is the greatest

amount of contaminant that can be present in drinking water without causing a risk to human health.

Mayordomo: A Spanish word that is used in the Southwest to identify the local person who is responsible for overseeing water allocation and maintenance of the water conveyance systems. A southwestern synonym for water commissioner or ditch rider.

Milligram (mg): One-thousandth of a grain.

Milligram per Liter (mg/l): A unit of the concentration of a constituent in water or wastewater. It Represents 0.001 gram of a constituent in 1 liter of water. It is approximately equal to one part per million (PPM).

Million Gallons per Day (MgD): A rate of flow of water equal to 133,680 cubic feet per day, or 1.572 cubic feet per second, or 3.0689-acre feet per day. A flow of one million gallons per day for one year equals 1,120 acre-feet (365 million gallons).

Mining Water Use: Water used during quarrying rocks and extracting materials from the land.

Municipal Water Systems: A water system that has at least five service connections or which regularly serves 25 individuals for 60 days; also called a public water system.

N

Navigable Waters: Those waters capable of supporting commerce.

Nephelometric Turbidity Unit (NTU): Unit of measure for the turbidity of water. Essentially, a measure of the cloudiness of water as measured by a nephelometer. Turbidity is based on the amount of light that is reflected off particles in the water.

Non-Consumptive Use: Use of water with return to the stream or watercourse of substantially the same amount of water as withdrawn; a use in which only insignificant amounts of water lost.

Nonpoint Source Pollution: Diffuse, overland runoff containing pollutants; pollution discharges from over a wide land area, not from one location.

O

Organic Matter: Plant and animal residues, or substances made by living organisms. All are based upon carbon compounds.

Osmosis: The movement of water molecules through a thin membrane. The osmosis process occurs in our bodies and is also one method of desalinating saline water.

Outfall: The place where a sewer, drain or stream discharges; the outlet or structure through which reclaimed water or treated effluent is finally discharged to a receiving water body.

Outwash: Stratified drift deposited by melt water streams beyond active glacier ice.

Oxygen Demand: The need for molecular oxygen to meet the needs of biological and chemical processes in water. Even though very little oxygen will dissolve in water, it is extremely important in biological and chemical processes.

P

Particle Size: the diameter, in millimeters, of suspended sediment or bed rock material. Particle size classifications are: (1) Clay – 0.00024-0.004 millimeters (mm) (2) Silt – 0.004-0.062 mm (3) Sand - 0.062-2.0 mm (4) Gravel – 2.0-640 mm

Parts per Billion: The number of "parts" by weight of a substance per billion parts of water. Used to measure extremely small concentrations.

Parts per Million: The number of "parts" by weight of a substance per million parts of water. Commonly used to represent pollutant concentrations.

Pathogen: A disease-producing agent; usually applied to a living organism. Generally, any viruses, bacteria, or fungi that cause disease.

Peak Flow: The maximum instantaneous discharge of a stream or river at a given location. It usually occurs at or near the time of maximum stage.

Per Capita Use: The average amount of water used per person during a standard time, generally per day.

Percolation/Infiltration: Downward flow or seepage of water through the pore and spaces of rock or soil.

Percolating Waters: Waters that pass through the ground beneath the surface of the earth without any definite channel and that do not form a part of the body or flow, surface or subterranean, of any watercourse. May be rainwater slowly infiltrating through soil or water seeping through banks or the bed of the stream, but these waters have left the flow of the stream so that they no longer may be characterized as a part of the stream flow.

Perfected Water Right: A completed or fully executed water right. A water right is said to have been perfected when all terms and conditions associated with it have been fully accomplished (e.g., a diversion has been completed and the water has been put to beneficial use).

Permeability: The capacity of a porous rock, sediment, or soil for transmitting fluid.

pH: A measure of the acidity or alkalinity of a solution, numerically equal to 7 for neutral solutions, increasing with alkalinity and decreasing with increasing acidity.

Photogeology: The geologic interpretation of aerial photographs. Th study and interpretation of photographs for the purpose of obtaining geologic and hydrologic information.

Piezometer: Casing and screen installed to measure the pressure head or water surface of an aquifer.

Point-Source Pollution: Water pollution coming from a single point, such as a sewage-outflow pipe.

Polychlorinated Biphenyls (PCBs): A group of synthetic, toxic industrial chemical compounds once used in making paint and electrical transformers, which are chemically inert and not biodegradable. PCBs were frequently found in industrial waters, and subsequently found their way into surface and ground waters. As a result of their persistence, they tend to accumulate in the environment. In terms of streams and rivers, PCBs are drawn to sediment, to which they attach and can remain virtually indefinitely. Although virtually banned in 1979 with the passage of the Toxic Substance Control Act, they continue to appear in the flesh of fish and other animals.

Porosity: The percentage of the total volume of a material that is open space.

Potable Water: Water of a quality suitable for drinking.

Potentiometric Surface: The level to which water rises in a well drilled into a confined aquifer.

Practicably Irrigable Acreage: A standard used to determine the extent of a federal reserved water right in Indian Reservations that was established to transform the Native Americans into farmers. The practicably irrigable acreage (PIA) standard was established in Arizona vs. California (1963). A more recent court case involving a general stream adjudication in the Big Horn Basin, Wyoming (1988), applied a two-part test to determine PIA lands, those irrigable at a reasonable cost.

Precipitation: Any form of rain, snow or hail falling to the earth's surface.

Preference System: State laws or constitutional provisions that establish a preference ranking for certain types of water use over others. Commonly rank domestic or municipal use as the highest use, agricultural use second, and industrial and mining uses third.

Pressure Energy: The energy given to groundwater by the weight of overlying water and earth materials.

Primary Wastewater Treatment: The first stage of the wastewater treatment process where mechanical methods, such as filters and scrapers are used to remove pollutants. Solid material in sewage also settles out in this process.

Prior Appropriation Doctrine: The western system of water appropriation that establishes water rights based on the priority of water use that is an individual's right to a specific quantity of water depends on when the use began. The first person to use the water from the source established the first right, the second person established the next right, and so on. During dry years, the person with the first right (senior right) has the first chance to use available water. The holder of the second right (a junior right) would have the second use, and so on.

S

Secondary Wastewater Treatment: Treatment (following primary wastewater treatment) involving the biological process of reducing suspended, colloidal, and dissolved organic matter in effluent from primary treatment systems and which generally removes 80 to 95 percent of the Biochemical Oxygen Demand (DOD) and suspended matter. Secondary wastewater treatment may be accomplished by biological or chemical physical methods. Activated sludge and trickling filters are two of the most common means of secondary treatment. It is accomplished by bringing together waste, bacteria, and oxygen in trickling filters or in the activated sludge process. This treatment removes floating and settleable solids and about 90 percent of the oxygen-demanding substances and suspended solids. Disinfections is the final stage of secondary treatment.

Sediment: Usually applied to material in suspension in water or recently deposited form suspension. In the plural the word is applied to all kinds of deposits from the waters of streams, lakes, or seas.

Sedimentary Rock: Rock formed of sediment and specifically: (1) Sandstone and shale, formed of fragments of other rock transported form their sources and deposited in water; and (2) rocks formed by or from secretions of organisms, such as most limestone. Many sedimentary rocks show distinct layering, which is the result of different types of sediment being deposited in succession.

Sedimentation Tanks: Wastewater tanks in which floating wastes are skimmed off and settled solids are removed for disposal.

Self-Supplied Water: Water withdrawn from a surface or groundwater source by a user rather than being obtained from a public supply. An example would b homeowners getting their water from their own well.

Seepage: (1) The slow movement of water through small cracks, pores, interstices, etc., of a material into or out of a body of surface or subsurface water. (2) The loss of water by infiltration into the soil from a canal, ditches, laterals, watercourse, reservoir, storage facilities or other body of water, or from a field.

Septic Tank: A tank used to drain domestic waster to allow the settling of solid prior to distribution to a leach field for soil absorption. Septic tanks are used when a sewer line is not available to carry them to a treatment plant. A settling tank in which settled sludge is in immediate contact with sewage flowing through the tank, and wherein solids are decomposed by anaerobic bacteria action.

Settling Pond (water quality): An open lagoon into which wastewater contaminated with solid pollutants is placed and allowed to stand. The solid pollutants suspended in the water sink to the bottom of the lagoon and the liquid can overflow out of the enclosure.

Severance: The act of severing or separating. In most western states, water rights are considered attached to the land on which they are used. When land is sold, water rights automatically transfer with the land to the new owner, unless such rights are specified severed from the land by deed.

Sewage Treatment Plant: A facility designed to receive the wastewater from domestic sources and to remove materials that damage water quality and threaten public health and safety when discharged into receiving streams or bodies of water. The substances removed are classified into four basic areas: (1) Greases and fats (2) Solids from human waste and other sources (3) Dissolved pollutants from human waste and decomposition products (4) Dangerous microorganisms Most facilities employ a combination of mechanical remove steps and bacterial decomposition to achieve the desired results. Chlorine is often added to discharges from the plants to reduce the danger of spreading disease by the release of pathogenic bacteria.

Sewer: A system of underground pipes that collect and deliver wastewater to treatment facilities or streams.

Sinkhole: A depression in the earth's surface caused by dissolving of underlying limestone, salt, or gypsum. Drainage is provided through underground channels that may be enlarged by the collapse of a cavern roof.

Solute: A substance that is dissolved in another substance, thus forming a solution.

Solution: A mixture of a solvent and a solute. In some solutions, such as sugar water, the substances mix so thoroughly that the solute cannot be seen. But in other solutions, such as water mixed with dye, the solution is visibly changed.

Spray Irrigation: A common irrigation method where water is shot from higher pressures sprayer onto crops. Because water is shot high into the air onto crops, some water is lost to evaporation.

Solvent: A substance that dissolves other substances, thus forming a solution. Water dissolves more substances than any other and is known as the "universal solvent".

Specific Capacity: The rate of discharge of a water borehole per unit drawdown, commonly expressed in gallons per minute per foot of drawdown.

Specific Heat: The number of calories required to raise the temperature of water one degree.

Spring: A place where groundwater flows naturally form a rock or the soil onto the land surface or into a body of surface water. Its occurrence depends on the nature and relationships of rocks, especially permeable and impermeable strata, on the position of the water table, and on the topography.

Spring Discharge: The water that issues the surface from an opening at land surface. Discharge is generally expressed as gallons per minute.

Spring Flow Meter: An instrument used to measure the flow from a spring.

Storage Coefficient: For an aquifer, the volume of water released from storage in a vertical column of 1.0 square foot when the water table or other piezometric surface declines 1.0 foot.

Storage Right: The authority from a responsible state entity to impound water in a reservoir.

Storm Right: The authority from a responsible state entity to impound water in a reservoir.

Storm Sewer: A sewer that carries only surface runoff, street wash, and snow melt from the land. In a separate sewer system, storm sewers are separate from those that carry domestic and commercial wastewater (sanitary sewers).

Stream: A general term for a body of flowing water; natural water course containing water at least part of the year. In hydrology, it is generally applied to the water flowing in a natural channel as distinct from a canal.

Streamflow: The water discharge that occurs in a natural channel. A more general term than runoff, streamflow may be applied to discharge whether it is affected by diversion or regulation.

Subsidence: A dropping of the land surface because of groundwater being pumped. Cracks and fissures can appear in the land. Subsidence is virtually an irreversible process.

Surface Tension: The attraction of molecules to each other on a liquid's surface. Thus, a barrier is created between the air and the liquid.

Surface Water: Water above the surface of the land, including (but not limited to) lakes, rivers, streams, diffused surface water, wastewater, floodwater, and ponds.

Suspended Sediment: Very fine soil particles that remain in suspension in water for a considerable period without contact with the bottom. Such material remains in suspension due to the upward components of turbulence and currents and/or by suspension.

Suspended-Sediments Concentration: the ratio of the mass of dry sediment in a water-sediment mixture. Typically expressed in milligrams of dry sediment per liter of water-sediment mixture.

Suspended- Sediment Discharge: The quantity of suspended sediment passing a point in a stream over a specified period. When expressed in tons per day, it is computed by multiplying water discharge (in cubic free per second) by the suspended-sediment concentration (in milligrams per liter) and by the factor 0.0027.

Suspended Solids: Solids that are not in true solutions and that can be removed by filtration. Such suspended solids usually contribute directly to turbidity. Defined in waste management, these are small particles of solid pollutants that resist separation by conventional methods.

T

Temporary Transfer: A transfer of water right from one purpose to another for a designated period.

Tertiary Wastewater Treatment: Selected biological, physical, and chemical separation processes to remove organic and inorganic substances that resist conventional treatment practices; the additional treatment of effluent beyond that of primary and secondary treatment methods to obtain a very high quality of effluent. The complete wastewater treatment process typically involves a three-phase process: (1) First, in the primary wastewater treatment process, which incorporates physical aspects, untreated water is passed through a series of screens to remove solid wastes; (2) Second, in the secondary wastewater treatment process, typically involving biological and chemical processes, screened wastewater is then passed a series of holding and aeration tanks and ponds; (3) Third, the tertiary wastewater treatment process consists of flocculation basins, clarifiers, filters, and chlorine basins or ozone or ultraviolet radiation processes.

Texture: The percentage of silt, sand and clay materials making up a soil.

Thermal Pollution: A reduction in water quality caused by increasing its temperature, often due to disposal of waste heat from industrial or power generation processes. Thermally polluted water can harm the environment because plants and animals can have a hard time adapting to it.

Thermoelectric Power Water Use: Water used in the process of the generation of thermoelectric power. Power plants that burn coal and oil are examples of thermoelectric-power facilities.

Trans-Basin: Typically, transportation of a water from one river basin to another river basin.

Transfer: The act of transferring water right from one person to another. Most states require that some formal notice or filing be made with an appropriate state agency so that the transaction is recorded officially, and the new owner is noted in the official water rights record.

Transmissivity: The rate at which water of the prevailing kinematic viscosity is transmitted through a unit width of the aquifer under a hydraulic gradient.

Transpiration: Process by which water that is absorbed by plants, usually through the roots, is evaporated into the atmosphere from the plant surface, such as leaf pores.

Tributary: A smaller river or stream that flows into a larger river or stream. Usually, several smaller tributaries merge to form a river.

Turbidity: The ability of a beam of light to pass through a sample of water.

U

Unconfined Aquifer: An aquifer that has no confining layer above it, so that the water table is exposed to atmospheric pressure and forms the upper boundary of the aquifer.

U.S. Geological Survey: An agency of the Federal Government of the Interior responsible for evaluating the mineral and water resources of the United States.

Unsaturated Zone: The zone between the land surface and the water table in which some of the pore spaces are filled with air and some are filled with water.

Upwelling: The rising of cold, heavy, subsurface water toward the surface.

Usufructuary Right: A water right holder's authority to divert and use a certain amount of water.

V

Vested Water Right: A fully executed or finalized appropriative right to use state water for a beneficial purpose.

W

Wastewater: Water that has been used in homes, industries and businesses that is not for reuse unless it is treated.

Wastewater: Treatment Return Flow – Water returned to the environment by wastewater- treatment facilities.

Water Administration: A broad term referring to the collective role of defined state agencies to implement state and federal water law, commonly though the development and implementation of appropriate regulations. This role can include oversight, approval, and enforcement responsibilities.

Water Bank: A mechanism for holding water for eventual use. Several western states use water banks of different varieties. For example, since the 1930's farmers in Idaho have marketed water through the Upper Snake Water Bank. Using three reservoirs (American Fall, Jackson Lake, and Palisades), irrigators have been able to maintain large water supplies as a hedge against drought. Another type of bank was established by the Kern County Water Agency of Central California with the California State Department of Water Resources. It involved purchase of 46,000 acres of land to develop an underground storage facility. During years of surplus, the underground water is sold by the state to parties who need it. A third example is the East Columbia Basin Water Bank operated by the East Columbia Basin Irrigation District. The bank leases water that have been salvaged because of conversation to sprinkler irrigation and changes in cropping patterns.

Water Commissioner: A person whose job is to make sure the water of each stream under his or her control is distributed in proper quantities at the right times to those who are authorized to receive it. May also be called a ditch rider or mayordomo. May be elected or hired by local water users or appointed by a state authority such as the state engineer or a judge.

Water Cycle: The circuit of water movement from the oceans to the atmosphere and to the Earth and return to the atmosphere through various stages or processes such as precipitation, interception, runoff, infiltration, percolation, storage, evaporation, and transportation.

Water Permit: A state license to appropriate water for a beneficial purpose.

Water Quality: A term used to describe the chemical, physical and biological characteristics of water, usually in respect to its suitability for a particular purpose.

Water Reservation: A water right granted by a state entity (commonly to public entities and on behalf of the public) for existing or future beneficial uses or for maintenance of a minimum flow, level, or quality of water.

Water Right: A legal right to divert state water for a beneficial purpose.

Watershed: The land area that drains water to a particular stream, river, or lake. It is a land feature that can be identified by tracing a line along the highest elevations between two area on a map, often a ridge. Large watersheds, like the Mississippi River basin contain thousands of smaller watersheds.

Water Table: The upper boundary of the zone of saturation.

Water Use: Water that is used for a specific purpose, such as for domestic use, irrigation or human interaction with and influence on the hydrologic cycle, and includes elements, such as water withdrawal form surface – and -groundwater sources, water delivery to homes and businesses, consumptive use of water, water released from wastewater treatment plants, water returned to the environment, and instream uses, such as using water to produce hydroelectric power.

Watthour (Wh): An electrical energy unit of measure equal to one watt of power supplied to or taken from an electrical circuit steadily for one hour.

Well (water): An artificial excavation put down by any method for the purposes of withdrawing water from the underground aquifers. A bored, drilled, or driven shaft or dug hole whose depth is greater than the largest surface dimensions and whose purpose is to reach underground water supplies or oil; or to store or bury fluids below ground.

Withdrawal: Water is removed from a groundwater or surface water source for use.

X

Xeriscaping: A method of landscaping that uses plants that are well adapted to the local area and are drought-resistant. Xeriscaping is becoming more popular as a way of saving water for domestic use.

Y

Yield: Mass per unit times per unit area.

Z

Zone of Aeration: Area between the ground surface and the water table.

Zone of Saturation: Area underground n which every available space is filled with water.

REFERENCES

Rampino, M.R., Calderia, K., & Zhu, Y. (2022). *A Pulse of the Earth: A 27.5 Myr underlying cycle in coordinated geological events over the last 260Myr.* doi:10.1016/j.gsf.2021.101245

U.S. Geological Survey. (2022). *Educational Resources.* https://www.usgs.gov/

Conclusion

"Tell me a fact and I'll learn. Tell me the truth and I'll believe. But tell me a story and it will live in my heart forever." – Indian Proverb

Stories are a sharing of knowledge passed down from our ancestors learned experiences. The power of telling stories cannot be underestimated. A story is a narrative account that raises unanswered questions, unresolved conflicts and how to solve issues and problems. A story line starts with the beginning, middle and end. Stories tell us about the human condition, and how our actions give way to consequences that mark the course of our lives. A story is rich and deep, it is about individuals and places. When others listen to the story, they participate in the action of the narrative (Green & Brock, 2000). This can change behaviors!

The power of stories has been a part of the human condition for all time. In Hollywood, storytelling is a multi-million-dollar business. Stories is the mode of thinking before our formal education begins. In addition, stories are a powerful structure for organizing and transmitting information and creating meaning in our lives and environments (Schank and Abelson, 1995). Individuals listening to stories react to them automatically, participating, in a sense, in the action of the narrative.

STORYTELLING IN TEACHING

As a teacher, I can capitalize on the narrative structure of research as a quest for knowledge. When you incorporate all these systems into the material for your course, student understanding and learning increases (Polichak and Gerrig, 2002). Science is the process of solving mysteries. Journal articles are often advised to make their content good stories, and psychologist often start by confronting a problem by listening to a persons' "story". Telling a story of how one's research began will motivate students to think of their own approaches to solving problems. It creates a "sense of discovery" (Green, 2004).

The characters of a story are an important element, they put the human element into the story. Students will focus and remember the characters and will be able to reconstruct the lessons learned from the story. Making the material personally relevant will lead to a greater ability to apply and practice what has been learned. Complex and ambiguous concepts may be hard to understand. A story may provide a way for students to apply learning. A narrative opening makes learning more straightforward and simpler. Thus, allowing students to relax and grasp a concrete example before moving into more theory and findings (Green, 2004).

Telling a story can lead to better rapport between the teacher and the student. By sharing experiences, it gives the student and class a new outlook, and makes the subject matter relevant. It brings the human side into the classroom. This opens discussion, providing a reciprocity that helps create a learning atmosphere where students are more willing to share their own experiences and opinions.

STORYTELLING AND BEHAVIOR CHANGE

We learn to tell stories so that we can share experiences and use our stories to present knowledge and information to the world. We influence others to change. Our thinking and attitudes are influenced by stories. It is an effective strategy in teaching positive behaviors, particularly when it comes to environmental, social and health issues. In the midst of a pandemic, one of the biggest motivators to influence people to vaccinate is to tell stories of those who have suffered.

In our society we are so polarized we tend to make everything political. I listen every day to people on both sides of the political spectrum that believe in their ideology so strongly, they cannot listen or question anything. This week, I read an article on China. I mentioned it to a friend at a pool party. He argued with me, did not ask why I mentioned what I did, just immediately discounted it. It was endorsed by both sides of the political leaders. All are concerned.

I realized if I had told a story to go along with this information, perhaps he would have been more open to what I had to say. I realized I did it to him as well, it is very difficult not to fall into a trap of not listening. Listening when a story or anything is said, is a precious gift you can give someone. A story is a medium to share information in a non-threatening influential mode.

Adults tend to learn a great deal of their own profession, but it always surprised me how uninformed our society is about many of our complex issues. Our media and entertainment industry has contributed to an anti-intellectual society. Ray Williams in his journal, Anti-Intellectualism and the "Dumbing Down" of America (2014), determined there is a growing trend of anti-intellectualism elitism in American culture. This starts with denying science, the arts, and humanities, which are being replaced by entertainment, self-righteousness, and deliberate gullibility.

There is an entire generation being dumbed down by social media and reality TV. Williams (2014) says we are creating a world of angry dummies who feel they have a right, the authority and the need not only to comment on everything, but to make sure their voice is heard above the rest, and to drag down any opposing views through personal attacks, loud repetition, and confrontation. This trend in now establishing itself in politics and business leadership. Moreover, it is enhanced by a declining education system. This is a sign we should stop social media, get off our phone, and listen to more stories. Go back to our roots!

I have only shared a few of my stories of this time in my life, but I hope it will influence teachers to tell more stories, influence and motivate others to learn about water and sustainability, and take the politics out of our science so that the environment, water, and sustainability are only based on scientific fact, and not emotional ideology. Most of all, I hope this book and the stories, change behaviors. We only have one mother earth, we can all work to take care of her.

REFERENCES

Green, M. C. (2004). *Story Telling in Teaching*. Psychological Science. https://www.psychologicalscience.org/observer/storytelling-in-teaching

Green, M. C., & Brock, T. C. (2000). The role of transportation in the persuasiveness of public narratives. *Journal of Personality and Social Psychology*, *79*, 401–421.

Polichak, J. W., & Gerrig, R. J. (2002). Get up and win: Participatory responses to narrative. In M. C. Green, J. J. Strange, & T. C. Brock (Eds.), *Narrative impact: Social and cognitive foundations* (pp. 71–96). Erlbaum.

Schank, R. C., & Abelson, R. P. (1995). Knowledge and memory: The real story. In R. S. Wyer Jr., (Ed.), *Advances in social cognition* (Vol. 8, pp. 1–85). Erlbaum.

Williams, R. (2014). Anti-Intellectualism and the "Dumbing Down" of America. *Psychology Today*. https://www.psychologytoday.com/blog/wiredsuccess/201407/anti-intellectualism-and-the-dumbing-down-america

About the Author

Kristin Joyce Tardif is an Associate Professor and Program Lead for the Organizational Leadership and Office Management Technology programs at the University of Arkansas Fort Smith. She teaches Organizational Leadership, serves on the university's standing committees on assessment, curriculum development, and faculty awards, and is the Faculty Adviser to the Student Veterans Organization, Salute Honor Society, and Leadership Human Resource Student Association. Dr. Tardif has lived and worked internationally, and has developed Canadian, Indian and Latin American partnerships, study abroad programs, as well as maymester programs. Her research focuses on two areas: Global Citizen Leadership, and the competencies to successfully participate in international trade and global assignments; Disruptive Creative Leadership, and igniting the creative economy. Dr. Tardif has a PhD in Organizational Leadership from The Chicago School of Professional Psychology. A Masters of Liberal Studies in Organizational Leadership from Fort Hays State University. Dr. Tardif was an executive with Nestle Foods in the corporate affairs marketing, and natural resources departments.

Index

E

F

G

H

I

S

T

U

V

W

www.ingramcontent.com/pod-product-compliance
Lightning Source LLC
LaVergne TN
LVHW080853300926
849162LV00003B/415

* 9 7 8 1 7 9 9 8 9 6 9 1 3 *